Math Challenge II-A
Geometry

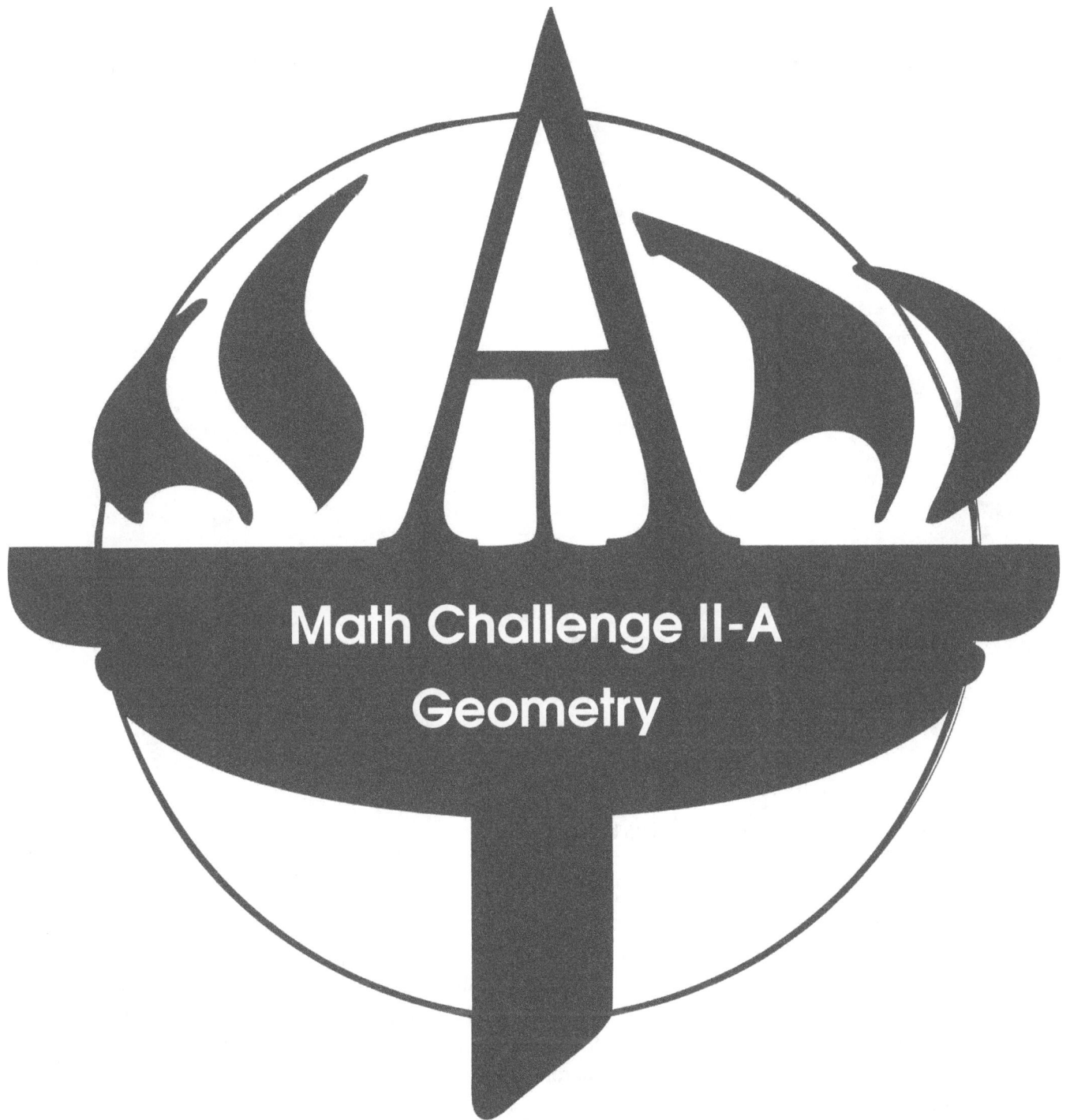

Areteem Institute

Math Challenge II-A Geometry

Edited by   John Lensmire
               David Reynoso
               Kevin Wang
               Kelly Ren

ISBN: 1-944863-18-4
ISBN-13: 978-1-944863-18-0
First printing, August 2018.

TITLES PUBLISHED BY ARETEEM PRESS

Cracking the High School Math Competitions (and Solutions Manual) - Covering AMC 10 & 12, ARML, and ZIML
Mathematical Wisdom in Everyday Life (and Solutions Manual) - From Common Core to Math Competitions
Geometry Problem Solving for Middle School (and Solutions Manual) - From Common Core to Math Competitions
Fun Math Problem Solving For Elementary School (and Solutions Manual)
ZIML Math Competition Book Division E 2016-2017
ZIML Math Competition Book Division M 2016-2017
ZIML Math Competition Book Division H 2016-2017
ZIML Math Competition Book Jr Varsity 2016-2017
ZIML Math Competition Book Varsity Division 2016-2017

MATH CHALLENGE CURRICULUM TEXTBOOK SERIES

Math Challenge I-A Pre-Algebra and Word Problems
Math Challenge I-B Pre-Algebra and Word Problems
Math Challenge I-C Algebra
Math Challenge II-A Algebra
Math Challenge II-B Algebra
Math Challenge III Algebra
Math Challenge I-A Geometry
Math Challenge I-B Geometry
Math Challenge I-C Topics in Algebra
Math Challenge II-A Geometry
Math Challenge II-B Geometry
Math Challenge III Geometry
Math Challenge I-B Counting and Probability
Math Challenge I-B Number Theory
Math Challenge II-A Number Theory

COMING SOON FROM ARETEEM PRESS

Fun Math Problem Solving For Elementary School Vol. 2 (and Solutions Manual)
Counting & Probability for Middle School (and Solutions Manual) - From Common Core to Math Competitions
Number Theory Problem Solving for Middle School (and Solutions Manual) - From Common Core to Math Competitions
Other volumes in the **Math Challenge Curriculum Textbook Series**

The books are available in paperback and eBook formats (including Kindle and other formats). To order the books, visit https://areteem.org/bookstore.

Contents

Introduction

The math challenge curriculum textbook series is designed to help students learn the fundamental mathematical concepts and practice their in-depth problem solving skills with selected exercise problems. Ideally, these textbooks are used together with Areteem Institute's corresponding courses, either taken as live classes or as self-paced classes. According to the experience levels of the students in mathematics, the following courses are offered:

- Fun Math Problem Solving for Elementary School (grades 3-5)
- Algebra Readiness (grade 5; preparing for middle school)
- Math Challenge I-A Series (grades 6-8; intro to problem solving)
- Math Challenge I-B Series (grades 6-8; intro to math contests e.g. AMC 8, ZIML Div M)
- Math Challenge I-C Series (grades 6-8; topics bridging middle and high schools)
- Math Challenge II-A Series (grades 9+ or younger students preparing for AMC 10)
- Math Challenge II-B Series (grades 9+ or younger students preparing for AMC 12)
- Math Challenge III Series (preparing for AIME, ZIML Varsity, or equivalent contests)
- Math Challenge IV Series (Math Olympiad level problem solving)

These courses are designed and developed by educational experts and industry professionals to bring real world applications into the STEM education. These programs are ideal for students who wish to win in Math Competitions (AMC, AIME, USAMO, IMO,

ARML, MathCounts, Math League, Math Olympiad, ZIML, etc.), Science Fairs (County Science Fairs, State Science Fairs, national programs like Intel Science and Engineering Fair, etc.) and Science Olympiad, or purely want to enrich their academic lives by taking more challenges and developing outstanding analytical, logical thinking and creative problem solving skills.

Math Challenge II-A is for students who are preparing for the American Mathematics Competition 10 (AMC 10) contest. Students are required to have fundamental knowledge in Algebra I, Geometry, Basic Number Theory and Counting and Probability up to the 10th grade level. Topics include polynomials, inequalities, special algebraic techniques, triangles and polygons, collinearity and concurrency, vectors and coordinates, numbers and divisibility, modular arithmetic, advanced counting strategies, binomial coefficients, sequence and series, and various other topics and problem solving techniques involved in math contests such as the AMC 10, advanced MathCounts, American Regions Math League (ARML), and Zoom International Math League (ZIML) Division H and Junior Varsity Division.

The course is divided into four terms:

- Summer, covering Algebra
- Fall, covering Geometry
- Winter, covering Combinatorics
- Spring, covering Number Theory

The book contains course materials for Math Challenge II-A: Geometry.

We recommend that students take all four terms. Each of the individual terms is self-contained and does not depend on other terms, so they do not need to be taken in order, and students can take single terms if they want to focus on specific topics.

Students can sign up for the course at `classes.areteem.org` for the live online version or at `edurila.com` for the self-paced version.

About Areteem Institute

Areteem Institute is an educational institution that develops and provides in-depth and advanced math and science programs for K-12 (Elementary School, Middle School, and High School) students and teachers. Areteem programs are accredited supplementary programs by the Western Association of Schools and Colleges (WASC). Students may attend the Areteem Institute in one or more of the following options:

- Live and real-time face-to-face online classes with audio, video, interactive online whiteboard, and text chatting capabilities;
- Self-paced classes by watching the recordings of the live classes;
- Short video courses for trending math, science, technology, engineering, English, and social studies topics;
- Summer Intensive Camps held on prestigious university campuses and Winter Boot Camps;
- Practice with selected free daily problems and monthly ZIML competitions at ziml.areteem.org.

Areteem courses are designed and developed by educational experts and industry professionals to bring real world applications into STEM education. The programs are ideal for students who wish to build their mathematical strength in order to excel academically and eventually win in Math Competitions (AMC, AIME, USAMO, IMO, ARML, MathCounts, Math Olympiad, ZIML, and other math leagues and tournaments, etc.), Science Fairs (County Science Fairs, State Science Fairs, national programs like Intel Science and Engineering Fair, etc.) and Science Olympiads, or for students who purely want to enrich their academic lives by taking more challenging courses and developing outstanding analytical, logical, and creative problem solving skills.

Since 2004 Areteem Institute has been teaching with methodology that is highly promoted by the new Common Core State Standards: stressing the conceptual level understanding of the math concepts, problem solving techniques, and solving problems with real world applications. With the guidance from experienced and passionate professors, students are motivated to explore concepts deeper by identifying an interesting problem, researching it, analyzing it, and using a critical thinking approach to come up with multiple solutions.

Thousands of math students who have been trained at Areteem have achieved top honors and earned top awards in major national and international math competitions, including Gold Medalists in the International Math Olympiad (IMO), top winners and qualifiers at the USA Math Olympiad (USAMO/JMO) and AIME, top winners at the

Zoom International Math League (ZIML), and top winners at the MathCounts National Competition. Many Areteem Alumni have graduated from high school and gone on to enter their dream colleges such as MIT, Cal Tech, Harvard, Stanford, Yale, Princeton, U Penn, Harvey Mudd College, UC Berkeley, or UCLA. Those who have graduated from colleges are now playing important roles in their fields of endeavor.

Further information about Areteem Institute, as well as updates and errata of this book, can be found online at http://www.areteem.org.

Acknowledgments

This book contains many years of collaborative work by the staff of Areteem Institute. This book could not have existed without their efforts. Huge thanks go to the Areteem staff for their contributions!

The examples and problems in this book were either created by the Areteem staff or adapted from various sources, including other books and online resources. Especially, some good problems from previous math competitions and contests such as AMC, AIME, ARML, MATHCOUNTS, and ZIML are chosen as examples to illustrate concepts or problem-solving techniques. The original resources are credited whenever possible. However, it is not practical to list all such resources. We extend our gratitude to the original authors of all these resources.

1. Special Angles I

Special Angles

- Special angles are the angles $30°$, $60°$, $90°$ and $45°$.
- An equilateral triangle has three $60°$ angles.
- In $\triangle ABC$, if $\angle C = 90°$, $\angle A = \angle B = 45°$, then $AB = \sqrt{2}AC = \sqrt{2}BC$.
- In $\triangle ABC$, if $\angle C = 90°$, $\angle A = 60°$, and $\angle B = 30°$, then $AB = 2AC$, $BC = \sqrt{3}AC$.

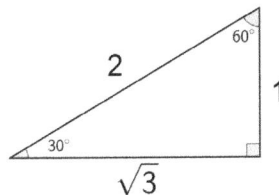

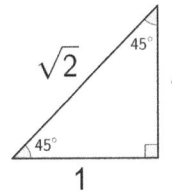

1.1 Example Questions

Problem 1.1 Suppose that *ABCD* is a square, and that *CDP* is an equilateral triangle, with *P outside* the square. What is the size of angle *PAD*?

Problem 1.2 Given square $ABCD$, let P and Q be the points outside the square that make triangles CDP and BCQ equilateral. Segments AQ and BP intersect at T. Find angle ATP.

Problem 1.3 Squares $OPAL$ and $KEPT$ are attached to the outside of equilateral triangle PEA. Draw segment TO, then find the size of angle TOP.

Problem 1.4 Mark P inside square $ABCD$, so that triangle ABP is equilateral. Let Q be the intersection of BP with diagonal AC. Triangle CPQ looks isosceles. Is this actually true?

Problem 1.5 Let triangle ABC be equilateral triangle with side length 16. Let D be on side $\overline{AB}$ and E be on side $\overline{AC}$ such that $\overline{DE} \| \overline{BC}$. Assume triangle ADE and trapezoid $DECB$ have the same perimeter. What is the length of $\overline{AD}$?

Problem 1.6 Let E be a point inside unit square $ABCD$ such that CDE is an equilateral triangle. Find the area of triangle AEC.

Problem 1.7 A triangle has a 60-degree angle and a 45-degree angle, and the side opposite the 45-degree angle has length 12. How long is the side opposite the 60-degree angle?

Problem 1.8 Let ABC be a triangle with $AB = AC$ and $\angle BAC = 20°$, and let P be a point on side AB such that $AP = BC$. Construct point D such that triangle ACD is equilateral, as shown in the diagram below. Show that triangle DCP is isosceles.

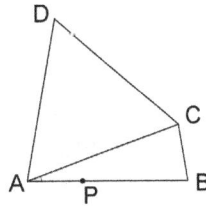

Problem 1.9 In triangle ABC, $AB = AC$, and BD is the altitude on AC. Given that $BD = \sqrt{3}$, and $\angle DBC = 60°$, find the area of $\triangle ABC$.

Problem 1.10 In a right triangle $\triangle ABC$ suppose $\angle B = 90°$ and $\angle C = 30°$. Suppose point D is on $\overline{BC}$ with $\angle ADB = 45°$ and $DC = 10$. Find the length of AB.

1.2 Quick Response Questions

Problem 1.11 If one of the leg lengths of a $45-45-90$ triangle is 3, the length of the hypotenuse is $A\sqrt{B}$. What is $A+B$?

Problem 1.12 If the shorter leg length of a $30-60-90$ triangle is 3, what is the length of the hypotenuse? Round your answer to the nearest tenth if necessary.

Problem 1.13 Consider the diagram below.

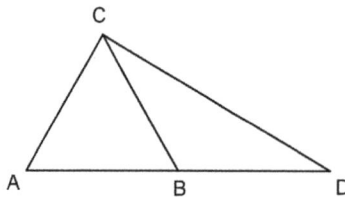

Suppose $AB=AC=1$ and $\angle BAC=60°$ and $\angle ADC=30°$. Find BD.

Problem 1.14 Suppose an isosceles triangle has an angle of $30°$. What is the measure of the largest angle the triangle may have?

Problem 1.15 Suppose you have an equilateral triangle. What are the angles of a right triangle created by drawing an altitude?

(A) $45-45-90$
(B) $30-60-90$
(C) $60-60-60$
(D) None of the above

Problem 1.16 Suppose you have an equilateral triangle and you form a right triangle by drawing an altitude. On the right triangle, how does the short side compare with the other two sides?

(A)   The short side is half the hypotenuse, and $1/\sqrt{3}$ of the other leg.
(B)   The short side is $1/\sqrt{3}$ of the hypotenuse and half of the other leg.
(C)   They are all the same size.
(D)   None of the above

Problem 1.17 The sides of an equilateral triangle are 4 cm long. An altitude of this triangle is $A\sqrt{B}$, what is $A + B$?

Problem 1.18 In the following diagram $\angle ADB = 90°$, $AD = 3$, $BD = 3$ and $BC = 6$. What is the measure of $\angle ABC$?

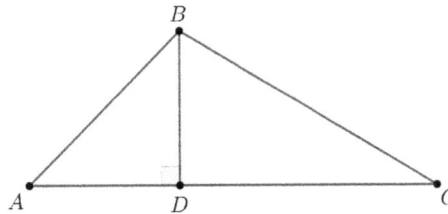

Problem 1.19 Suppose you have a trapezoid $ABCD$, with AB paralel to CD, $AD = 5\sqrt{2}$, $BC = 5\sqrt{2}$, $AB = 20$ and $CD = 10$. What is the area of the trapezoid?

Problem 1.20 On the following diagram $BD = 3$ and $AC = P + \sqrt{Q}$. What is $P + Q$?

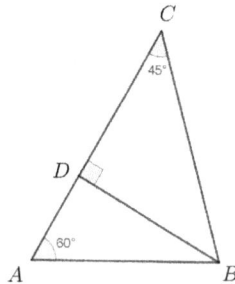

1.3 Practice Questions

Problem 1.21 Suppose that *ABCD* is a square, and that *CDP* is an equilateral triangle, with *P inside* the square.

(a) What is the size of $\angle PAD$?

(b) What is the size of $\angle BPA$?

Problem 1.22 Given square *ABCD*, let *P* and *Q* be the points outside the square that make triangles *CDP* and *BCQ* equilateral. Show triangle *APQ* is equilateral.

Problem 1.23 Squares *OPAL* and *KEPT* are attached to the outside of equilateral triangle *PEA*. Decide whether segments *EO* and *AK* have the same length, and give your reasons.

Problem 1.24 Mark *P* inside square *ABCD*, so that triangle *ABP* is equilateral. Find the size of $\angle CPD$.

Problem 1.25 Let triangle *ABC* be equilateral triangle. Let *D* be on side $\overline{AB}$ and *E* be on side $\overline{AC}$ such that $\overline{DE}\|\overline{BC}$. Let the perimeter of triangle *ADE* to be 12 and the perimeter of trapezoid *DECB* to be 16. Find the perimeter of triangle *ABC*.

Problem 1.26 Let *E* be a point outside unit square *ABCD* such that *CDE* is an equilateral triangle. Find the area of triangle *AEC*.

Problem 1.27 A triangle has a 45-degree angle and a 30-degree angle, and the side opposite the 45-degree angle has length 12. How long is the side opposite the 30-degree angle?

Problem 1.28 Let ABC be a triangle with $AB = AC$ and $\angle BAC = 20°$, and let P be a point on side AB such that $AP = BC$. Find $\angle ACP$.

Problem 1.29 Given that one of the angles of the triangle with sides $(5, 7, 8)$ is $60°$, show that one of the angles of the triangle with sides $(3, 5, 7)$ is $120°$.

Problem 1.30 In a right triangle $\triangle ABC$ suppose $\angle B = 90°$ and $\angle A = 45°$. Suppose point D is on $\overline{BC}$ with $\angle ADB = 60°$ and $DC = 5$. Find the length of AB.

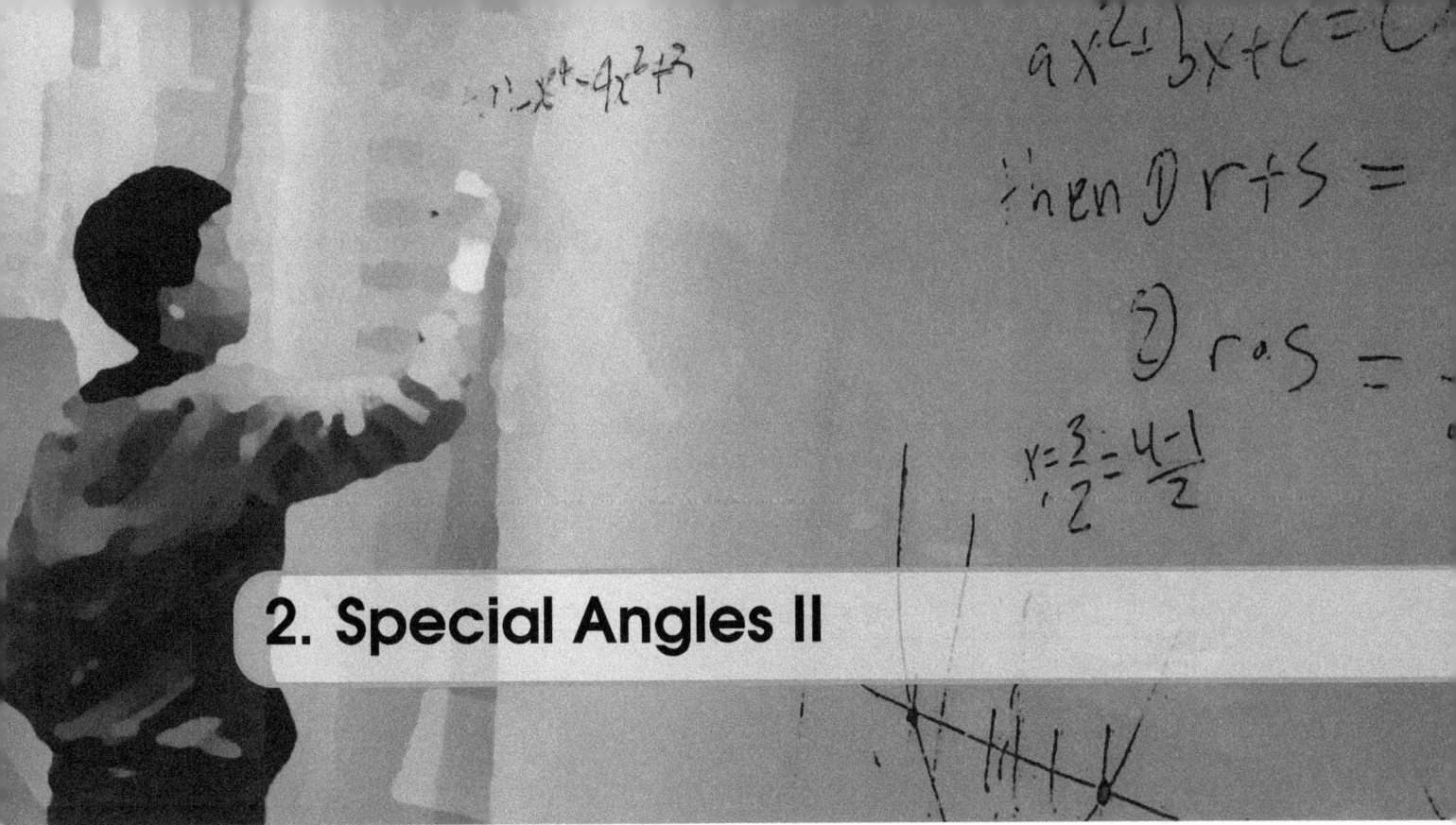

2. Special Angles II

Review and Definitions

- **Equilateral polygon:** A polygon that has all of its sides of the same length is called equilateral.
- **Equiangular polygon:** A polygon that has all of its internal angles the same size is called equiangular.
- **Regular polygon:** A polygon that is both equilateral and equiangular is called regular.
- We use $[ABC]$ to denote the area of triangle ABC, $[DEFG]$ to denote the area of quadrilateral $DEFG$, etc.

2.1 Example Questions

Problem 2.1 Complete the following table about polygons with *n* sides: name, sum of interior angles, sum of exterior angles, and measure of each angle in case of regular polygon. All angles are in degrees. Justify your answers. Keep the chart for your own reference.

n	Name	Int. Angle Sum	Ext. Angle Sum	Each Angle (if regular)
3	Triangle			
4				
5				
6				
7	Heptagon			
8				
9	Nonagon			
10				
12	Dodecagon			
20	Icosagon			

Problem 2.2 Use 6 equilateral triangles to form a hexagon *ABCDEF*.

(a) Show hexagon $ABCDEF$ is regular. Justify your answer.

(b) Calculate the angle AED.

Problem 2.3 Four non-overlapping regular plane polygons all have sides of length 1. The polygons meet at a point A in such a way that the sum of the four interior angles at A is $360°$. Among the four polygons, two are squares and one is a triangle. What is the last polygon?

Problem 2.4 Find the area of the largest equilateral triangle that fits in a regular hexagon of area 50.

Problem 2.5 In equiangular octagon $ABCDEFGH$, $AB = CD = EF = GH = 6\sqrt{2}$ and $BC = DE = FG = HA$. Given the area of the octagon is 184, compute the length of side BC.

Problem 2.6 Answer the following

(a) Given a 60-120 isosceles trapezoid, prove that the sum of the length of the top and one of the side is equal to the length of the base.

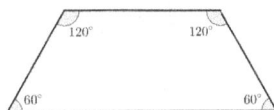

(b) Let triangle ABC be a equilateral triangle with length equals to 1. Let point P be the center of the triangle ABC, D be point on $\overline{BC}$, E be point on $\overline{CA}$, and F be point on $\overline{AB}$, so that $\overline{PD}\|\overline{AB}$, $\overline{PE}\|\overline{BC}$, and $\overline{PF}\|\overline{AC}$. Find the value of $PD+PE+PF$.

Problem 2.7 Show that a regular dodecagon (12-sided polygon) can be cut into pieces that are all regular polygons, which need not all have the same number of sides.

Problem 2.8 What is the side length of the largest equilateral triangle that can fit inside a 2-by-2 square?

Problem 2.9 Let $ABCD$ be a unit square and let P, Q be on sides $\overline{AD}, \overline{AB}$ respectively such that $\triangle APQ$ has perimeter 2. Rotate $\triangle PDC$ $90°$ about C. Call the point P is rotated to P'. Prove that $\triangle PQC$ is congruent to $\triangle P'QC$.

Problem 2.10 In convex quadrilateral $ABCD$, $\angle A = 60°$, $\angle C = 30°$, and $AB = AD$. Show that $AC^2 = BC^2 + CD^2$.

2.2 Quick Response Questions

Problem 2.11 Is the following a regular polygon?

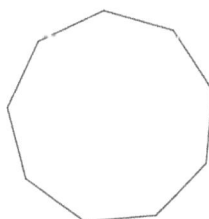

Problem 2.12 Is the following an equiangular polygon that is not equilateral?

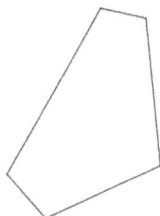

Problem 2.13 Is the following an equilateral polygon that is not equiangular?

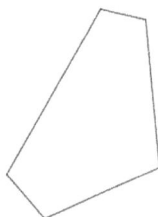

Problem 2.14 Let $ABCDEF$ be a regular hexagon, and let $EFGHI$ be a regular pentagon. $\angle GAF$ could have measure α or β degrees, where $\alpha > \beta$. Find $\alpha - \beta$.

Problem 2.15 A regular n-sided polygon has exterior angles of m degrees each. What is m?

(A) $m = 360$
(B) $m = 180(n-2)$
(C) $m = \frac{360}{n}$
(D) $m = 360(n-2)$

Problem 2.16 Let $\triangle ABC$ be an isosceles triangle with $\angle A = 50°$. What is the sum of all possible angle measures of $\angle B$?

Problem 2.17 A regular n-sided polygon has exterior angles of m degrees each. For how many polygons is m a whole number?

Problem 2.18 Consider a regular nonagon $COMPUTERS$. Is it true that $TE + ES = SP$?

Problem 2.19 As shown in the diagram below, the equiangular convex hexagon $ABCDEF$ has $AB = 1$, $BC = 4$, $CD = 2$, and $DE = 4$. $[ABCDEF] = \frac{P\sqrt{Q}}{R}$, with $\gcd(P.R) = 1$ and such that R has no square factors. What is $P + Q + R$?

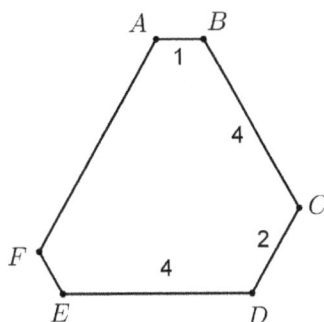

Problem 2.20 Find the area of the largest regular hexagon that fits in an equilateral triangle of area 108.

2.3 Practice Questions

Problem 2.21 Give two more examples of n so that a regular polygon with n sides has integer interior angles.

Problem 2.22 Use 6 equilateral triangles to form a hexagon $ABCDEF$, then find the area of triangle AED in terms of the total area.

Problem 2.23 Four non-overlapping regular plane polygons all have sides of length 1. The polygons meet at a point A in such a way that the sum of the four interior angles at A is $360°$. Among the four polygons, two are squares and one is a triangle. What is the perimeter of the entire shape?

Problem 2.24 Construct the largest equilateral triangle in a regular hexagon, and find the perimeter of the triangle given that the side lengths of the hexagon are 4.

Problem 2.25 In equiangular octagon $ABCDEFGH$, $AB = CD = EF = GH$ and $BC = DE = FG = HA$. Argue that $\overline{AB} \perp \overline{CD}$.

Problem 2.26 Let triangle ABC be a equilateral triangle with length equals to 1. Let point P be an arbitrary point in the interior of triangle ABC, D be point on $\overline{BC}$, E be point on $\overline{CA}$, and F be point on $\overline{AB}$, so that $\overline{PD}\|\overline{AB}$, $\overline{PE}\|\overline{BC}$, and $\overline{PF}\|\overline{AC}$. Find the value of $PD + PE + PF$.

Problem 2.27 Come up with a pattern to cover the entire plane (called a *tiling*) with regular hexagons, squares, and triangles (each with the same side length). Do this in a way such that no two hexagons are adjacent to each other.

Problem 2.28 What is the area of the *smallest* equilateral triangle that can be inscribed in a 1-by-1 square? Note: Inscribed here means all 3 vertices of the triangle are on the square.

Problem 2.29 Let $ABCD$ be a unit square and let P, Q be on sides $\overline{AD}, \overline{AB}$ respectively such that $\triangle APQ$ has perimeter 2. Find the measure of $\angle PCQ$.

Problem 2.30 Give an example of a convex quadrilateral $ABCD$ (complete with the measure of all angles and all sides) that satisfies the constraints: $\angle A = 60°$, $\angle C = 30°$, and $AB = AD$. (Recall we have shown that $AC^2 = BC^2 + CD^2$.)

3. Analytical Geometry

Introduction

- Analytic geometry, also called coordinate geometry, or sometimes algebraic geometry, is the study of geometry using a coordinate system.
- For notation purposes, given a point A, write its coordinates in the xy-plane as (x_A, y_A).

Beginning Formulas

- Distance Formula: The length AB is given by $\sqrt{(x_A - x_B)^2 + (y_A - y_B)^2}$.
- Midpoint Formula: The midpoint of $\overline{AB}$ is $\left(\dfrac{x_A + x_B}{2}, \dfrac{y_A + y_B}{2} \right)$.
- Generalized Midpoint Formula: Let $A \neq B$ be two points and C be a point between A, B on $\overline{AB}$. Then if $\dfrac{AC}{AB} = \lambda$ (so λ is between 0 and 1), then
$$C = ((1 - \lambda)x_A + \lambda x_B, (1 - \lambda)y_A + \lambda y_B).$$

Slopes

- Slope between two points A and B: $\dfrac{y_A - y_B}{x_A - x_B}$.
- Horizontal Lines have slope 0.
- Vertical Lines have undefined or infinite slope.
- Parallel Lines have the same slope.
- Perpendicular Lines have opposite reciprocal slopes. That is, if you multiply the slopes together you get -1.

Lines

- Standard form: $Ax + By = C$.
- Slope-intercept form: $y = mx + b$ (m is the slope, b is the y-intercept).
- Point-slope form: $y - y_A = m(x - x_A)$ for a line with slope m going through point A.

Beginning Conics

- Translations:
    - Replacing x by $x - h$ in an equation shifts the graph h units right.
    - Replacing y by $y - k$ in an equation shifts the graph k units up.

Circles

- $(x - h)^2 + (y - k)^2 = r^2$ where (h, k) is the center of the circle and r is the radius.

Parabolas

- $y - k = a(x - h)^2$ for a parabola opening vertically with vertex (h, k).
- $x - k = a(y - h)^2$ for a parabola opening horizontally with vertex (h, k).

3.1 Example Questions

Problem 3.1 (Not analytic geometry) Prove that in a right triangle $\triangle ABC$ (with $\angle B = 90°$), that the perpendicular bisectors of $\overline{AB}$ and $\overline{BC}$ meet at the midpoint of AC.

Problem 3.2 Prove that the equation for the line going through points A and B is given by

$$\frac{x - x_A}{x_A - x_B} = \frac{y - y_A}{y_A - y_B}.$$

Problem 3.3 Suppose a triangle has vertices $(3,4), (4,7), (7,6)$. Find the area of the triangle. Hint: One possible method is to find the altitude from $(4,7)$.

Problem 3.4 Suppose $\triangle ABC$ is an equilateral triangle with with $A = (0,2)$ and $B = (\sqrt{3},1)$. Find all possible coordinates for C.

Problem 3.5 Suppose you have a line $\ell : Ax + By + C = 0$ and a point P.

(a) If $A = 0$ find the (shortest) distance from P to ℓ.

(b) If $B = 0$ find the (shortest) distance from P to ℓ.

(c) The general equation for the (shortest) distance d from P to ℓ is

$$d = \frac{|A \cdot x_P + B \cdot y_P + C|}{\sqrt{A^2 + B^2}}.$$

Verify this equation for the line $3x + 4y + 6 = 0$ and $P = (8, -5)$. That is, calculate the distance WITHOUT using the formula, and check your work with the formula.

Problem 3.6 Prove that quadrilateral $ABCD$ is a parallelogram if and only if

$$\begin{cases} x_A + x_C &= x_B + x_D, \\ y_A + y_C &= y_B + y_D. \end{cases}$$

Problem 3.7 Circles

(a) Prove (not necessarily using analytic geometry) that if a circle has diameter $\overline{AB}$, then a point P lies on the circle if and only if $\angle APB = 90°$.

(b) Prove that if $\overline{AB}$ is the diameter of a circle, then the circle has equation $(x - x_A)(x - x_B) + (y - y_A)(y - y_B) = 0$.

Problem 3.8 One definition of a parabola is the collection of point that are equidistant from a point F (called the focus) and a line d (called the directrix). Verify that the parabola $y = \frac{1}{4}x^2$ consists of the points equidistant from $F : (0, 1)$ and $d : y = -1$.

Problem 3.9 Find the shortest path starting and ending at the origin that goes around the circle $(x - 4)^2 + y^2 = 8$.

Problem 3.10 A parabola has equation $y = x^2 + bx + c$ and the line $y = 5$ intersects the parabola at $x = -1, 3$.

(a) Find b and c.

(b) Find the vertex of the parabola.

(c) Find the focus and directrix of the parabola.

3.2 Quick Response Questions

Problem 3.11 Suppose for a line we measure the angle from the x-axis to the line (counterclockwise). Therefore $0°$ is a horizontal line and $90°$ is a vertical line. What slope corresponds to an angle of $45°$?

(A) 1

(B) $\dfrac{\sqrt{3}}{3}$

(C) $\sqrt{3}$

(D) None of the above

Problem 3.12 Suppose for a line we measure the angle from the x-axis to the line (counterclockwise). Therefore $0°$ is a horizontal line and $90°$ is a vertical line. What slope corresponds to an angle of $60°$?

(A) 1

(B) $\dfrac{\sqrt{3}}{3}$

(C) $\sqrt{3}$

(D) None of the above

Problem 3.13 Suppose slope m has angle θ, where $0° < \theta < 180°$. What angle does slope $-m$ have?

(A) 2θ

(B) $180° + \theta$

(C) $180° - \theta$

(D) None of the above

Problem 3.14 The distance from the line $y = 2x + 5$ to the origin is $\sqrt{A}$. What is A? Hint: You can use formulas from class.

Problem 3.15 The point (a,b) over the line $y = 2x + 5$ is the closest to the origin. What is $a + b$?

Problem 3.16 At how many points does the parabola $y = 2x^2 - 8x + 10$ intersect the circle $x^2 + y^2 = 4x$? Hint: Try completing the square!

Problem 3.17 What is the radius of the circle given by the equation $x^2 + 4x + (y - 1)^2 = 21$?

Problem 3.18 The line given by $\dfrac{x-4}{2} = \dfrac{y-5}{13}$ goes through points $A = (4,5)$ and B. What is y_B?

Problem 3.19 $ABCD$ is a parallelogram with $A = (-2,4)$ and $B = (2,5)$. What are be the coordinates of C and D?

(A) $C = (-2,4)$ and $D = (2,5)$
(B) $C = (2,5)$ and $D = (2,-4)$
(C) $C = (0,4)$ and $D = (0,5)$
(D) $C = (0,4)$ and $D = (-4,3)$

Problem 3.20 The parabola $y = \frac{1}{32}x^2$ has focus $F = (0,8)$ and directrix $y = k$. What is k?

3.3 Practice Questions

Problem 3.21 Let $A = (-1, 1), B = (0, 0), C = (1, 1)$. Verify that the perpendicular bisectors of $\overline{AB}$ and $\overline{BC}$ meet at the midpoint of $\overline{AC}$.

Problem 3.22 Suppose that a, b are respectively the x-intercept and y-intercept of a line. Show that the line has equation $\dfrac{x}{a} + \dfrac{y}{b} = 1$.

Problem 3.23 Suppose you start at the origin and walk along the line $y = x$ until you reach the line $3y = 30 - 2x$. You then walk along this second line until you reach the x-axis again. Finally you return to the origin along the x-axis.

(a) How far have you walked?

(b) What is the area of the region you enclosed during your walk?

Problem 3.24 Suppose $\triangle ABC$ is an isosceles triangle with with $A = (0, 2)$ and $B = (2, 1)$. Find all possible coordinates for C.

Problem 3.25 Consider lines $\ell_1 : y = x - 2, \ell_2 : y = x + 2, \ell_3 : y = 6 - x$. Find the largest circle such that (i) The center of the circle is on line ℓ_3 and (ii) The circle does not go outside line ℓ_1 and ℓ_2.

Problem 3.26 Suppose it is given that $ABCD$ is a parallelogram if and only if

$$\begin{cases} x_A + x_C &= x_B + x_D, \\ y_A + y_C &= y_B + y_D. \end{cases}$$

Come up with and prove a rule for showing when quadrilateral $ABCD$ is a rectangle. It may not be as simple as the rule before, but try to make it as easy as possible.

Problem 3.27 Suppose you have a circle $(x-1)^2 + (y-1)^2 = 4$ and a line $x + y = 4$.

(a) The line and circle intersect at two points A, B, find them.

(b) Verify that the perpendicular bisector of $\overline{AB}$ goes through the center of the circle.

(c) Find the area of $\triangle ABC$ where C is the center of the circle.

Problem 3.28 Find the equation for a parabola with focus $(0, p)$ and directrix $y = -p$.

Problem 3.29 Find the equation for the line with slope 1 that is tangent to the upper half of the circle $(x-3)^2 + (y-1)^2 = 2$.

Problem 3.30 Find the focus and directrix of the parabola $2y = 4 - 2x - x^2$.

4. Areas I

Concepts and Facts

- **Notation:** We shall use $[ABC]$ to denote the area of triangle ABC, $[XYZW]$ to denote the area of the quadrilateral $XYZW$, etc.
- Formulas for areas (should be memorized): square, rectangle, triangle, parallelogram, trapezoid, circle

  Various area formulas of triangle ABC:

$$
\begin{aligned}
[ABC] &= \frac{1}{2}ah && (h \text{ is the altitude on } a)\\
&= \frac{abc}{4R} && (R \text{ is the circumradius})\\
&= \sqrt{s(s-a)(s-b)(s-c)} && (s = \frac{a+b+c}{2})\\
&= rs && (r \text{ is the inradius, } s \text{ is defined as above})
\end{aligned}
$$

- The areas of triangles (or parallelograms) with equal bases and equal altitudes (heights) are equal.
- The areas of triangles with equal altitudes are proportional to the bases of the triangles.
- The ratio of areas between two similar triangles is the square of the ratio between the corresponding sides.

4.1 Example Questions

Problem 4.1 Using only the basics about parallel lines and congruent/similar triangles and the fact that the area of a rectangle is bh, prove the following. (Note: once a fact is proven below, you can use it in later parts.)

(a) The area of a parallelogram is bh.

(b) The area of a triangle is $\frac{1}{2}bh$ (prove this two ways!).

(c) The area of a trapezoid is $\frac{b_1+b_2}{2}h$.

(d) The area of a trapezoid is also mh where m is the *median* of the trapezoid, which connects the midpoints of the two non-parallel sides.

Problem 4.2 What is the ratio of areas between two similar triangles? Prove your result!

Problem 4.3 Prove the Pythagorean Theorem using areas.

Problem 4.4 In $\triangle ABC$, $AB > AC > BC$, $\overline{CD}, \overline{BE}, \overline{AF}$ are altitudes on $\overline{AB}, \overline{AC}, \overline{BC}$, respectively. Show that $CD < BE < AF$.

Problem 4.5 In triangle ABC, $AC = 10$, $BC = 24$, $AB = 26$. What is the altitude on $\overline{AB}$?

Problem 4.6 Let $ABCD$ be a parallelogram, and E, H, F, G be points on sides $\overline{AB}, \overline{BC}, \overline{CD}, \overline{DA}$ respectively, and $\overline{EF} \| \overline{BC}$ and $\overline{GH} \| \overline{AB}$. Let P be the intersection of $\overline{EF}$ and $\overline{GH}$. If $[GPFD] = 10, [PHCF] = 8, [EBHP] = 16$, find $[ABCD]$.

Problem 4.7 Suppose you only know that the centroid exists. Prove (using areas!) that the centroid divides each median in a ratio of $1 : 2$.

Problem 4.8 Let $ABCD$ be a parallelogram, with midpoints E, F, G, H (say on $\overline{AB}, \overline{BC}, \overline{CD}, \overline{DA}$). Let I, J be the midpoints of $\overline{EF}, \overline{GH}$. Find the area of $\triangle JIG$ as a fraction of the area of $ABCD$.

Problem 4.9 Prove that if a triangle has side lengths a, b, c, inradius r, and circumradius R we have $2Rr = \dfrac{abc}{a+b+c}$.

Problem 4.10 Let $ABCD$ be a parallelogram as in the diagram, with E the midpoint of $\overline{BC}$.

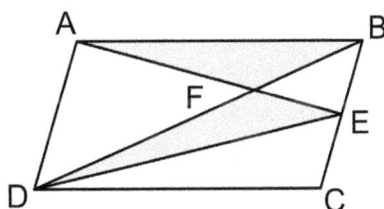

(a) Compare the shaded regions $\triangle ABF$ and $\triangle DEF$, which one has the larger area?

(b) Find the area of the shaded regions $\triangle ABF$ and $\triangle DEF$ in terms of the entire area of the parallelogram.

4.2 Quick Response Questions

Problem 4.11 If $AB \parallel CD$, can we conclude that $[ABC] = [ABD]$?

Problem 4.12 Consider the two squares in the diagram below.

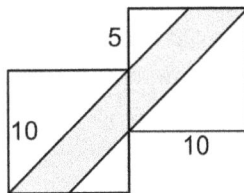

Find the area of the shaded region.

Problem 4.13 Consider two squares attached at their corner vertices as in the diagram below.

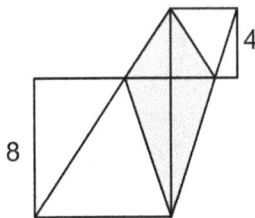

Find the area of the shaded region.

Problem 4.14 Suppose $\triangle ABC$ with E on $\overline{AB}$ and D on $\overline{AC}$ such that $AE = AB/3, AD = AC/2$. If $[AED] = 2$, find the area of $[ABC]$.

Problem 4.15 In triangle ABC, $AC = 9$, $BC = 10$, $AB = 17$. What is the altitude on $\overline{BC}$?

Problem 4.16 $\triangle ABC$ has area 54. Let D be a point on AB such that $AD = 2DB$. What is $[ADC]$?

Problem 4.17 Triangle ABC is inscribed on a circle of radius 10. $AB = 19$, $AC = 12$, and $[ABC] = 65.25$. What is BC? Round your answer to the nearest hundredth.

Problem 4.18 Let $\triangle ABC$ be an isosceles triangle with $AC = BC$. Let D, E and F be the midpoints of BC, CA and AB, respectively and let G be its centroid. If $EB = 12$, what is AG?

Problem 4.19 Consider the parallelogram $ABCD$. Let E and F be midpoints of the sides CD and DA, respectively. If $[ABCD] = 40$, what is $[BEF]$?

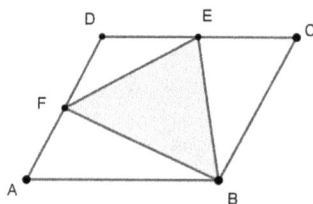

Problem 4.20 In the following diagram, G is the centroid of $\triangle ABC$. If $[ABC] = 92$, what is the shaded area?

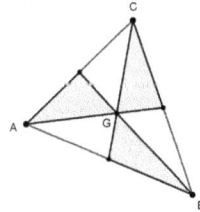

4.3 Practice Questions

Problem 4.21 Recall that a kite is a quadrilateral with two sets of adjacent sides of equal length.

(a) Prove that the diagonals in a kite are perpendicular.

(b) Prove that the area of a kite is $\dfrac{d_1 \cdot d_2}{2}$ (where d_1, d_2 are the diagonals) using the same ideas as Problem 4.1.

Problem 4.22 In $\triangle ABC$, let D, E, F be midpoints of the sides $\overline{BC}, \overline{AC}, \overline{AB}$. Show that $[DEF] = [ABC]/4$.

Problem 4.23 Prove the converse of the Pythagorean Theorem. Hint: You can use the Pythagorean Theorem!

Problem 4.24 Let $ABCD$ be a parallelogram with $[ABCD] = 1$. Let P be a point in the interior of $ABCD$. Show that $[ABP] + [CDP]$ is a fixed value and find that value.

Problem 4.25 Find a formula for the area of a parallelogram if you are given the two sides lengths as well as one of the diagonals.

Problem 4.26 Let $ABCD$ be a rectangle, and E, H, F, G be points on sides $\overline{AB}, \overline{BC}, \overline{CD}, \overline{DA}$ respectively, and $\overline{EF} \| \overline{BC}$ and $\overline{GH} \| \overline{AB}$. Let P be the intersection of EF and GH. If $[GPFD] = 10, [EBHP] = 12$ and $[ABCD] = 44$. Suppose each of the four rectangles formed above has integer dimensions, find all possible dimensions for $GPFD$ and $EBHP$.

Problem 4.27 Prove that the centroid is well-defined using areas (that is, show the medians are concurrent). Hint: Your proof will be an extension of the one done in class.

Problem 4.28 In parallelogram $ABCD$, M and N are midpoints of sides $\overline{AB}$ and $\overline{BC}$ respectively. Given that $[DMN] = 9$, find the area of $ABCD$.

Problem 4.29 Prove that if a triangle has side lengths a, b, c semiperimeter s, inradius r, and circumradius R we have

$$\frac{R}{r} = \frac{abc}{4(s-a)(s-b)(s-c)}.$$

Problem 4.30 Suppose a parallelogram P_1 has area 256. Connect the midpoints of each side to form a parallelogram P_2. Repeat to get P_3, and continue repeating until you get to P_{10}.

(a) Prove this problem actually makes sense. That is, prove that if you connect the midpoints of a parallelogram you get another parallelogram.

(b) Find the area of P_{10}.

5. Centers of Triangles

Definitions

- **Concurrency:** Three distinct lines (or segments) are *concurrent* if they pass through the same point.
- **Collinearity:** Three distinct points are *collinear* if they lie on the same straight line.
- **Cevian:** In a triangle, the line segment joining a vertex to any given point on the opposite side (or line through opposite side) is called a *cevian*.
- **Median:** The line from a vertex to the midpoint on the opposite side.
- **Altitude:** The line from a vertex which is perpendicular to the opposite side.
- **Angle Bisector:** The line from a vertex which bisects the angle at the vertex.

Centers of Triangles and Theorems about Concurrency

- The three medians of a triangle are concurrent. This point is called the *centroid* of the triangle.
- The three angle bisectors of a triangle are concurrent. This point is called the *incenter* of the triangle.
- The three altitudes of a triangle are concurrent. This point is called the *orthocenter* of the triangle.
- The three perpendicular bisectors of the sides of a triangle are concurrent. This point is called the *circumcenter* of the triangle.

Angle Bisector Theorem

- In a triangle $\triangle ABC$, let D be on $\overline{BC}$ so that $\overline{AD}$ is the angle bisector of $\angle A$. Then $\dfrac{AB}{AC} = \dfrac{BD}{DC}$.

5.1 Example Questions

Problem 5.1 Prove that the perpendicular bisectors are concurrent (and therefore, the circumcenter is well-defined).

Problem 5.2 Prove that the altitudes of a triangle are concurrent (and therefore, the orthocenter is well-defined). Use the previous result and the following hint: Given triangle $\triangle ABC$ construct lines through each vertex parallel to the opposite side to form $\triangle PQR$.

Problem 5.3 Prove the Angle Bisector Theorem using the following hint: In $\triangle ABC$, let E be on ray $\overrightarrow{AD}$ such that $\overline{CE} \parallel \overline{AB}$.

Problem 5.4 Suppose a circle is drawn outside a triangle, so that the circle passes through each vertex of the triangle (the circle is *circumscribed* on the triangle). Find, with proof, the center of this circle. Can you guess another name for this circle?

Problem 5.5 Suppose $\triangle ABC$ is a right triangle with $\angle C = 90°$ and $AC = 5, BC = 12$. Let $\overline{AD}$ be an angle bisector. Find the area of $\triangle ABD$.

Problem 5.6 Viviani's Theorem

(a) Recall a more general version of the problem we did earlier: Let triangle ABC be a equilateral triangle with side length s. Let point P be an arbitrary point in the interior of triangle ABC, D be point on $\overline{BC}$, E be point on $\overline{CA}$, and F be point on $\overline{AB}$, so that $\overline{PD}\|\overline{AB}$, $\overline{PE}\|\overline{BC}$, and $\overline{PF}\|\overline{AC}$. Find the value of $PD+PE+PF$.

(b) (Viviani's Theorem) Let triangle ABC be an equilateral triangle with side length s and therefore heights/altitudes of length $h = \dfrac{s\sqrt{3}}{2}$. Let P be an arbitrary point in the interior of the triangle. Show that the sum of the distances from P to all three sides is equal to h.

Problem 5.7 Prove that the coordinates of the centroid of $\triangle ABC$ are given by

$$\left(\frac{x_A+x_B+x_C}{3}, \frac{y_A+y_B+y_C}{3}\right).$$

Further prove that the centroid divides the medians into ratios of $2:1$.

Problem 5.8 Let quadrilateral $ABCD$ be in the diagram below, AE and AF are angle bisectors.

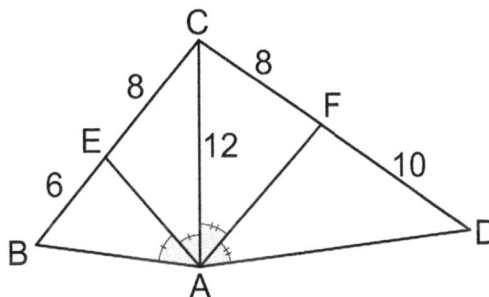

Find the perimeter of $ABCD$.

Problem 5.9 Suppose three congruent circles are all tangent inside a larger circle.

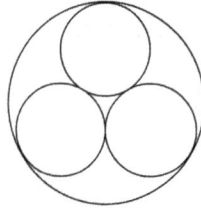

If the bottom two circles have centers $(2,0)$ and $(4,0)$ find the equation of the large circle.

Problem 5.10 (Euler Line) Let ABC be a triangle with centroid G and circumcenter O.

(a) Let H be on ray $\overrightarrow{OG}$ so that $OG : GH = 1 : 2$. Prove that $\overline{AH} \perp \overline{BC}$.

(b) Prove that the centroid, circumcenter, and orthocenter of $\triangle ABC$ are collinear. If this line is unique it is called the Euler line.

5.2 Quick Response Questions

Problem 5.11 Which of the following are always inside the triangle?

(A) Incenter and Cirumcenter
(B) Incenter and Orthocenter
(C) Orthocenter and Circumcenter
(D) Centroid and Incenter

Problem 5.12 When is the Circumcenter inside of the triangle?

(A) When the triangle is obtuse
(B) When the triangle is acute
(C) When the triangle is right
(D) Always

Problem 5.13 When is the Orthocenter outside of the triangle?

(A) When the triangle is obtuse
(B) When the triangle is acute
(C) When the triangle is right
(D) Always

Problem 5.14 For which of the following triangles is it always true that the Incenter and Circumcenter are the same?

(A) Isosceles triangle
(B) Equilateral triangle
(C) Right triangle
(D) None of the above

Problem 5.15 (Golden Triangle) In $\triangle ABC$, $AB = AC$. Point D is on side $\overline{AB}$ such that $\overline{CD}$ bisects $\angle ACB$, and $CD = BC$. What is the measure of $\angle A$?

Problem 5.16 Let $\triangle ABC$ with $AB = 4$, $BC = 3$, $AC = 5$. Let BE be the angle bisector of B. CE has length $\frac{P}{Q}$ as a reduced fraction. What is $P + Q$?

Problem 5.17 Let $\triangle ABC$ with $AB = 4$, $BC = 3$, $AC = 5$. Let BE be the angle bisector of B. The area of $\triangle CBE$ is $\frac{P}{Q}$. What is $P + Q$?

Problem 5.18 Consider $\triangle ABC$ and a point D on side CA such that $\angle ABD = \angle CBD$. If $AD = 4$, $CD = 5$ and $AB = 8$, what is BC?

Problem 5.19 Consider points $A = (-4, 9)$, $B = (-6, 4)$ and $C = (4, 8)$. The coordinates of the centroid of $\triangle ABC$ are (a, b). What is $a + b$?

Problem 5.20 Let $\triangle ABC$ be an equilateral triangle and let P be a point in the interior of $\triangle ABC$ such that the sum of the distances from P to each of the sides is $8\sqrt{3}$. What is the side length of the triangle?

5.3 Practice Questions

Problem 5.21 Prove that the angle bisectors of a triangle are concurrent (and therefore, the incenter is well-defined).

Problem 5.22 Complete the steps below to prove (using classical geometry) that the medians of a triangle are concurrent (and therefore, the centroid is well-defined). Further, show that the centroid divides the medians into a ratio of $2 : 1$.

(a) Given a triangle $\triangle ABC$, let D, E be the midpoints of $\overline{AC}, \overline{BC}$ respectively. Prove that $DE = AB/2$ and $\overline{AB} \| \overline{DE}$.

(b) Let F be the intersection of the medians $\overline{BD}$ and $\overline{AE}$. Show that $AF : FE = BF : FD = 2 : 1$.

(c) Complete the entire proof.

Problem 5.23 Prove the converse of the Angle Bisector Theorem: Suppose $\triangle ABC$ is a triangle and D is on $\overline{BC}$ such that $\dfrac{AB}{AC} = \dfrac{BD}{DC}$. Prove that $\overline{AD}$ is the angle bisector of $\angle A$. Hint: Proceed similarly to the proof of the Angle Bisector Theorem.

Problem 5.24 Inscribed Circle

(a) Suppose a circle is drawn inside a triangle, so that the circle is tangent to all the sides of the triangle (the circle is inscribed in the triangle). Find, with proof, the center of this circle. Can you guess another name for this circle?

(b) Call the radius of the circle above the inradius and denote it r. If $s = \dfrac{a+b+c}{2}$ is the semiperimeter of the triangle, show that the area of the triangle is $r \cdot s$.

Problem 5.25 Suppose $\triangle ABC$ is a non-degenerate triangle with integer side lengths and let $\overline{AD}$ be an angle bisector. If $BD = 4, DC = 5$, find the smallest possible values for AB, AC.

Problem 5.26 The radius of the circumcircle is called the circumradius, denoted R. The area of a triangle (with side lengths a, b, c) is given by $\dfrac{abc}{4R}$.

(a) Prove this formula for a right triangle.

(b) Prove this formula for an equilateral triangle.

Problem 5.27 Suppose a triangle has vertices $A = (a, 0), B = (0, b), C = (0, c)$. (For help visualizing, pretend a is negative while b, c are positive. However, your work should not depend on this!)

(a) Prove (analytically) that the altitudes of $\triangle ABC$ all meet at a single point.

(b) Explain how your work above really proves at the altitudes of any triangle all meet in a single point.

Problem 5.28 Let $\triangle ABC$ have perimeter 35. Let AE be the angle bisector of $\angle A$, with $BE = 4$ and $CE = 10$. Find the length of AB and AC.

Problem 5.29 Find an equation of the circle with the smallest radius containing the circles $C_1 : (x-1)^2 + (y-1)^2 = 4$, $C_2 : (x-5)^2 + (y-1)^2 = 4$, $C_3 : (x-1)^2 + (y-5)^2 = 4$, $C_4 : (x-5)^2 + (y-5)^2 = 4$.

Problem 5.30 Recall that in a $\triangle ABC$, the centroid, circumcenter, and orthocenter are collinear as proved in Problem 5.10 If this line was unique we called it the Euler line.

(a) Explain why the Euler line is not defined for an equilateral triangle.

(b) Prove that in an isosceles triangle, the Euler line contains a vertex of the triangle.

6. Areas II

Concepts and Facts

- **Notation:** We shall use $[ABC]$ to denote the area of triangle ABC, $[XYZW]$ to denote the area of the quadrilateral $XYZW$, etc.
- Formulas for areas (should be memorized): square, rectangle, triangle, parallelogram, trapezoid, circle
  Various area formulas of triangle ABC:

$$
\begin{aligned}
[ABC] &= \frac{1}{2}ah & &(h \text{ is the altitude on } a) \\[2mm]
&= \frac{abc}{4R} & &(R \text{ is the circumradius}) \\[2mm]
&= \sqrt{s(s-a)(s-b)(s-c)} & &(s = \frac{a+b+c}{2}) \\[2mm]
&= rs & &(r \text{ is the inradius, } s \text{ is defined as above})
\end{aligned}
$$

- The areas of triangles (or parallelograms) with equal bases and equal altitudes (heights) are equal.
- The areas of triangles with equal altitudes are proportional to the bases of the triangles.
- The ratio of areas between two similar triangles is the square of the ratio between the corresponding sides.

6.1 Example Questions

Problem 6.1 Suppose you have a trapezoid $ABCD$ with $\overline{AB}$ parallel to $\overline{CD}$. Let E be the intersection of the diagonals. Suppose $AB = 10, CD = 15$ and $\triangle ADE$ has area 24. Find the area of $ABCD$.

Problem 6.2 Let $\overline{AM}$ be a median of $\triangle ABC$, and D be a point on $\overline{MC}$, and E be a point on $\overline{AB}$, such that $\overline{ME} \parallel \overline{AD}$. Show that $[BDE] = [AEDC]$.

Problem 6.3 Suppose the altitudes of a triangle are in ratio $2 : 2 : 3$ and the triangle has a perimeter of 24. Find the area of the triangle.

Problem 6.4 Let G be the centroid of $\triangle ABC$, and $AG = 3, BG = 4, CG = 5$, find $[ABC]$.

Problem 6.5 Let M be the intersection of line segments $\overline{AB}$ and $\overline{PQ}$. Show that
$$\frac{[PAB]}{[QAB]} = \frac{PM}{QM}.$$

Problem 6.6 (2002 AMC 12A #22) Triangle ABC is a right triangle with $\angle ACB$ as its right angle, $m\angle ABC = 60°$, and $AB = 10$. Let P be randomly chosen inside $\triangle ABC$, and extend $\overline{BP}$ to meet $\overline{AC}$ at D. What is the probability that $BD > 5\sqrt{2}$?

Problem 6.7 Suppose a dodecagon is in a square as in the diagram below (the vertices of the dodecagon on the square are the midpoints of the sides):

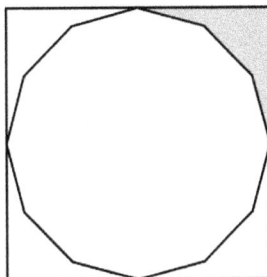

If the area of the square is 4, what is the area of the shaded region?

Problem 6.8 Let $ABCDE$ be a convex pentagon. Suppose further that the triangle cut off by each diagonal has area 1. What is the area of the full pentagon $ABCDE$?

Problem 6.9 Start with trapezoid $ABCD$. Extend $\overline{AD}, \overline{BC}$ to meet at O as in the diagram below.

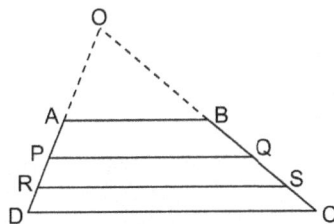

Suppose $AD = 1, OA = 1$. Construct P, Q, R, S such that $\overline{AB} \parallel \overline{PQ} \parallel \overline{RS}$ and $[ABQP] = [QSRP] = [SCDR]$. What are AP, PR, RD?

Problem 6.10 (Kurrah's Theorem) Let $\triangle ABC$ be given. Construct D, E as in the diagram below with $\angle ABC = \angle ADB = \angle BEC$.

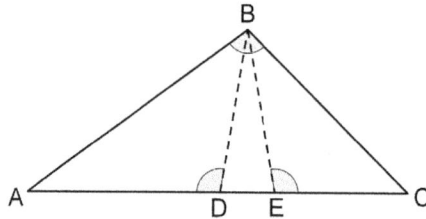

Prove that

$$AB^2 + BC^2 = AC(AD + CE).$$

6.2 Quick Response Questions

Problem 6.11 Suppose $\triangle ABC$ with E on $\overline{AB}$ and D on $\overline{AC}$ such that $AE = AB/3, AD = AC/2$. If $[AED] = 2$, find the area of $[ABC]$.

Problem 6.12 In the diagram below, there are 21 grid points arranged in equilateral triangles, equally spaced. The area of each small equilateral triangle formed by 3 adjacent grid points is 1. Find the area of $\triangle ABC$.

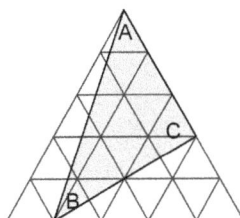

Problem 6.13 Suppose $ABCD$ is a parallelogram and P is an arbitrary point on the interior of $ABCD$. Which of the following is true?

  (A) $[ABP] = [ABCD]/4$
  (B) $[ABP] = [CDP]$
  (C) $[ABP] + [CDP] = [ABCD]/2$
  (D) None of the above

Problem 6.14 Let P be an interior point in parallelogram $ABCD$, and $[APB] : [ABCD] = 2 : 5$. $[CPD] : [ABCD] = a : b$, where a and b have no common factors. What is $a + b$?

Problem 6.15 Suppose you have a circle with diameter $\overline{AB}$ with $AB = 4$. Let C, D be on arc $\widehat{AB}$ such that $\widehat{AC} : \widehat{CD} : \widehat{DB} = 1 : 2 : 1$. Find the area of the figure enclosed by line segment $\overline{AC}$, arc $\widehat{CD}$, and line segment $\overline{AD}$. Use $\pi = 3.14$ and round your answer to the nearest tenth if necessary.

Problem 6.16 In $\triangle ABC$, $AB = 7$, $BC = 10$ and $AC = 14$. If $\angle ABC = \angle ADB = \angle CEB$, $DE = \frac{P}{Q}$ in lowest terms. What is $P + Q$?

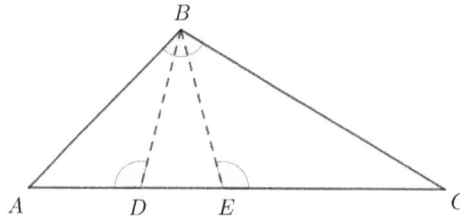

Problem 6.17 Suppose a dodecagon is in a square as in the diagram below (the vertices of the dodecagon on the square are the midpoints of the sides):

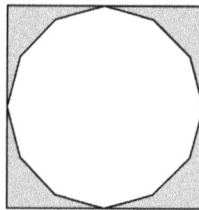

If the area of the square is 8, what is the area of the shaded region?

Problem 6.18 A triangle has side lengths 6, 8 and 12. The area of the triangle is $\sqrt{A}$, what A?

Problem 6.19 Consider the parallelogram $ABCD$. Let E be the middle point of AB and F be the middle point of CD. If $[ABCD] = 248$, what is $[EBF]$?

Problem 6.20 Let M be the intersection of the line segments $\overline{AB}$ and $\overline{PQ}$. If $PM : QM = 2 : 5$ and $[PAB] = 16$, what is $[QAB]$?

6.3 Practice Questions

Problem 6.21 Let $ABCD$ be a trapezoid with $\overline{AB}\|\overline{CD}$. Let E be the intersection of the two diagonals $\overline{AC}, \overline{BD}$.

(a) If $ABCD$ is a parallelogram show $[ABE] = [BCE] = [CDE] = [DAE]$.

(b) Prove that for a general trapezoid, $[BCE] = [DAE]$.

Problem 6.22 Given square $ABCD$, let E, F be the midpoints of AB and BC respectively, and G be the intersection of AF and CE. If $[ABCD] = 1$, find $[AGCD]$.

Problem 6.23 Suppose that a triangle has altitudes with ratio $12 : 15 : 20$ and area 192. Find the perimeter of the triangle.

Problem 6.24 ABC is a triangle with integer side lengths. Extend $\overline{AC}$ beyond C to point D such that $CD = 120$. Similarly, extend $\overline{CB}$ beyond B to point E such that $BE = 112$ and $\overline{BA}$ beyond A to point F such that $AF = 104$. If triangles CBD, BAE, and ACF all have the same area, what is the minimum possible area of triangle ABC?

Problem 6.25 Let O be the intersection of the diagonals of convex quadrilateral $ABCD$. Given that $[ABC] = 5, [ACD] = 10$, and $[ABD] = 6$, find $[ABO]$.

Problem 6.26 (2002 AMC 12A #23) In triangle ABC, side $\overline{AC}$ and the perpendicular bisector of $\overline{BC}$ meet in point D, and $\overline{BD}$ bisects $\angle ABC$. If $AD = 9$ and $DC = 7$, what is the area of triangle ABD?

Problem 6.27 What is the area of a dodecagon that is inscribed in a circle of radius 1?

Problem 6.28 Suppose $ABCD$ is a trapezoid with height 8. If the diagonals are perpendicular and one diagonal has length 10, what is the area of the trapezoid?

Problem 6.29 Let $\triangle ABC$. Construct $\overline{DE} \parallel \overline{FG} \parallel \overline{BC}$ (with D on $\overline{AB}$ and F on $\overline{DB}$) dividing the triangle into three regions of the same area. What is $AD : DF : FB$?

Problem 6.30 Explain why the Pythagorean theorem follows almost immediately from Kurrah's theorem.

7. Circles I

Basic Definitions

- A *circle* is a collection of points of equal distance (called the *radius*) from a set point (called the *center*).
- Given two points A, B on a circle, the segment $\overline{AB}$ is called a *chord*.
- If a chord AB contains the center of the circle, we say A and B are *diametrically opposite*, and call AB a *diameter*.
- The portion of a circle that lies above or below a chord AB is called an *arc*. If the arc is more than half a circle it is called a *major arc*, less than half a circle is called a *minor arc*, and half a circle is called a *semicircle*. The arc will be denoted $\overgroup{AB}$.
- Suppose $\overgroup{AB}$ is an arc on a circle with center O. The *angular size* of the arc $\overgroup{AB}$ is equal to the angle $\angle AOB$ (which is referred to as a *central angle*).
- Given a central angle $\angle AOB$ from an arc $\overgroup{AB}$, the figure contained between the arc $\overgroup{AB}$ and the radii $\overline{OA}, \overline{OB}$ is called a *sector*.

Measurements in Circles

- The area of a circle is given by πr^2 where r is the radius.
- The circumference of a circle is given by $2\pi r = \pi d$ where r, d are the radius and length of a diameter respectively.
- The *arc length* of $\overgroup{AB}$ (that is, the distance walking from A to B along the circle) is given by $\dfrac{\theta}{360°} 2\pi r$ where θ is the angular size of $\overgroup{AB}$ (measured in degrees).

- Similarly, the area of a sector from arc $\overset{\frown}{AB}$ is given by $\dfrac{\theta}{360°}\pi r^2$ where θ is the angular size of $\overset{\frown}{AB}$ (measured in degrees).

Theorems about Perpendiculars (Results will be proven below.)

- In a circle, a radius is perpendicular to a chord if and only if the radius bisects the chord.
- In a circle, the perpendicular bisector of a chord passes through the center of the chord.
- Similar to the above: Suppose a line going through a point P on the circle. The line is tangent to the circle if and only if the line is perpendicular to the radius of the circle.

Arcs and Angles (Results will be proven below.)

- If points A, B, P are on a circle, we call $\angle APB$ an *inscribed angle*. The measure of $\angle APB$ is half the angular size of arc $\overset{\frown}{AB}$ (where the arc does NOT contain P).
- Suppose two chords AC, BD intersect inside the circle at a point P as in the diagram below.

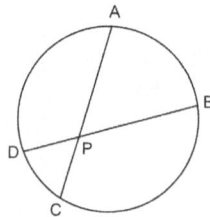

Then $\angle APB$ is half the sum of the angular sizes of arcs $\overset{\frown}{AB}$ and $\overset{\frown}{CD}$.
- Suppose the extension of two chords AC, BD intersect outside the circle at a point P as in the diagram below.

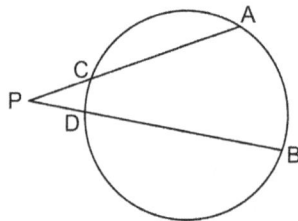

Then $\angle APB$ is half the difference of the angular sizes of arcs $\overset{\frown}{AB}$ and $\overset{\frown}{CD}$.

7.1 Example Questions

Problem 7.1 Storing Tires

(a) Suppose two tires, each with radius 1ft rest upright on the ground and touching each other, as pictured below:

How much space is needed horizontally to store the tires?

(b) Repeat part (a) with two tires of radius $1, 2$ feet respectively.

Problem 7.2 Suppose you start with a circle of radius 1. Draw another circle of radius 1 with center an arbitrary point on the first circle. Let R denote the region consisting of all points that are inside both circles.

(a) Find the perimeter of R.

(b) Find the area of R.

Problem 7.3 Inscribing Circles in Sectors

(a) What is the radius of the largest circle that can fit in a quarter circle of radius 1?

(b) What if instead you are fitting it into a 60°-sector?

Problem 7.4 Let $\angle APB$ be an inscribed angle on a circle with center O. Prove that $\angle APB$ is half the angular size of arc $\widehat{AB}$ if:

(a) O lies on $\angle APB$.

(b) O lies inside $\angle APB$.

Problem 7.5 Prove that if two chords AC, BD intersect inside a circle at point P then the measure of $\angle APB$ is half the sum of the angular sizes of $\widehat{AB}, \widehat{CD}$.

Problem 7.6 Suppose $\widehat{AB}$ is an arc with angular size 60° and CD is a diameter such that if rays $\overrightarrow{BA}, \overrightarrow{DC}$ are extended to intersect at a point E, $\angle AEC = 30$. Find the angular size of arc $\widehat{BD}$.

Problem 7.7 Suppose two perpendicular chords intersect and divide each other in a ratio of $1 : 2$. Find the radius of the circle if each chord is 12in long.

Problem 7.8 Suppose ω is a circle with radius 6 and center O. Let $\widehat{AB} = 135°$. Let C be on ω such that $\overline{OA} \parallel \overline{BC}$. Find $[OACB]$.

Problem 7.9 (AMC 12A 2007 #10) A triangle with side length in the ratio $3 : 4 : 5$ is inscribed in a circle of radius 3. What is the area of the triangle?

Problem 7.10 Let $\mathscr{C}_1$ and $\mathscr{C}_2$ be circles defined by

$$(x-10)^2 + y^2 = 36$$

and

$$(x+15)^2 + y^2 = 81,$$

respectively. What is the length of the shortest line segment $\overline{PQ}$ that is tangent to $\mathscr{C}_1$ at P and to $\mathscr{C}_2$ at Q?

7.2 Quick Response Questions

Problem 7.11 The circle $x^2 + y^2 + 10x - 24y - 87 = 0$ has center (h, k). What is $h + k$?

Problem 7.12 What is the radius of the circle $x^2 + y^2 + 10x - 24y - 87 = 0$?

Problem 7.13 If four circles are drawn in a plane, what is the maximum number of points that belong to at least two of the circles?

Problem 7.14 Arrange 4 congruent circles so that (i) the center of the four circles form a square with side length 10, and (ii) adjacent circles are tangent. What is the radius of each circle?

Problem 7.15 Arrange 4 congruent circles so that (i) the center of the four circles form a square with side length 10, and (ii) adjacent circles are tangent. Find the area of the region inside the square that is outside each of the circles. Use $\pi = 3.14$ and round your answer to the nearest tenth if necessary.

Problem 7.16 Suppose A, B, C are points on a circle such that the angular measures of arc $\widehat{AB}$ (not containing C) and arc $\widehat{CA}$ (not containing B) are in ratio $5 : 9$. Suppose further that $\angle ABC = 90°$. Find the measure of $\angle BAC$.

Problem 7.17 Suppose ω is a circle with radius 6 and center O. Let A, B and C be points on ω such that $\overline{OA} \parallel \overline{BC}$. What angle does $\widehat{AB}$ have to be so that $OACB$ is a parallelogram?

Problem 7.18 In the following diagram $\angle ABC = 42°$. What is the measure of $\angle AOC$?

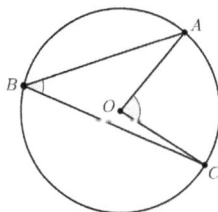

Problem 7.19 In the following diagram $\angle AOD = 100°$ and $\angle BPC = 66°$. What is the measure of angle $\angle BOC$?

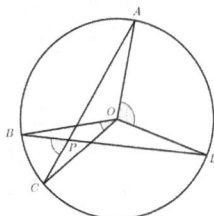

Problem 7.20 In the following diagram $\angle AOD = 115°$ and $\angle BOC = 23°$. What is the measure of angle $\angle BPC$?

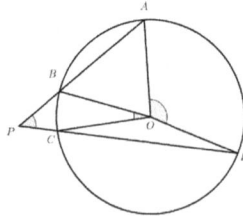

7.3 Practice Questions

Problem 7.21 Prove that in a circle, a radius is perpendicular to a chord if and only if the radius bisects the chord.

Problem 7.22 Prove that in a circle, the perpendicular bisector of a chord passes through the center of the circle.

Problem 7.23 Suppose you have three tires with radii $1, 2, 3$ feet respectively. You store them horizontally as in Problem 7.1 above. What is the minimum amount of horizontal space needed to store the tires? Justify your answer!

Problem 7.24 Suppose you start with a circle of radius 1. Draw a second circle of radius 1 centered at an arbitrary point on the first circle. Now pick one of the two intersection points and draw a third circle centered at this point, also with radius 1. Call R the region consisting of all points that are contained inside at least two circles.

(a) Find the perimeter of R.

(b) Find the area of R.

Problem 7.25 Inscribing Circles in Sectors

(a) What is the radius of the largest circle that can fit in a $120°$ sector of a circle of radius 1?

(b) What if you have a 240° sector?

Problem 7.26 Let $\angle APB$ be an inscribed angle on a circle with center O. Prove that $\angle APB$ is half the angular measure of arc $\widehat{AB}$ if O lies outside $\angle APB$.

Problem 7.27 Prove that if two chords AC, BD intersect outside a circle at point P then the measure of $\angle APB$ is half the difference of the angular sizes of $\widehat{AB}, \widehat{CD}$.

Problem 7.28 Suppose $\overline{AB}, \overline{CD}$ are two chords of equal length who intersect at E. Suppose $\angle AED = 120°$, and $AE : EB = CE : ED = 1 : 2$. Further, suppose $AC = 2$.

(a) Find the distance from E to the center of the circle.

(b) Find the radius of the circle.

Problem 7.29 Find the area of a 30-60-90 triangle circumscribed about a circle of radius 1.

Problem 7.30 Let $\mathscr{C}_1$ and $\mathscr{C}_2$ be circles defined by

$$(x-5)^2 + y^2 = 9$$

and

$$(x+5)^2 + y^2 = 9,$$

respectively. What is the length of the shortest line segment $\overline{PQ}$ that is tangent to $\mathscr{C}_1$ at P and to $\mathscr{C}_2$ at Q? Hint: There are a couple possible tangents.

8. Circles II

Cyclic Quadrilateral (Results Proven Below)

- Points A, B, C, D lie on circle ω in (clockwise) order. (We say quadrilateral $ABCD$ is *cyclic*.) Then $\angle A + \angle C = \angle B + \angle D = 180°$.
- Let $ABCD$ be a convex cyclic quadrilateral. Then $\angle DBA = \angle DCA, \angle ACB = \angle ADB, \angle BDC = \angle BAC$, and $\angle CBD = \angle CAD$.

Power of a Point

- Let $\overline{AB}, \overline{CD}$ be chords, which intersect at E inside the circle. Then $AE \cdot BE = CE \cdot DE$.
- Let $\overline{AB}, \overline{CD}$ be chords, which are extended to intersect at E outside the circle. Then $AE \cdot BE = CE \cdot DE$.
- Let a line tangent to the circle at C intersect the extension of chord $\overline{AB}$ at E. Then $AE \cdot BE = CE^2$. (Note: Think of this as the previous case with $C = D$.)

Ptolemy's Theorem (Results Proven Below)

- In a cyclic quadrilateral $ABCD$,

$$AC \cdot BD = AB \cdot CD + AD \cdot BC.$$

(In other words, if all the four vertices of a quadrilateral are on the same circle, then the product of the diagonals equals the sum of the products of the two pairs of opposite sides.)

8.1 Example Questions

Problem 8.1 Prove that if $\angle A + \angle C = \angle B + \angle D = 180°$ in quadrilateral $ABCD$, then $ABCD$ is a cyclic quadrilateral .

Problem 8.2 Suppose $ABCD$ is a quadrilateral.

(a) Prove that if $ABCD$ is cyclic, then $\angle ABD = \angle ACD$ (and similarly $\angle BAC = \angle BDC$, etc.).

(b) Prove that if $\angle ABD = \angle ACD$ (or $\angle BAC = \angle BDC$, etc.), then $ABCD$ is cyclic.

Problem 8.3 Prove the Power of a Point formula for $\overline{AB}$ and $\overline{CD}$ intersecting inside the circle.

Problem 8.4 Prove Ptolemy's theorem, using the following diagram as guidance,

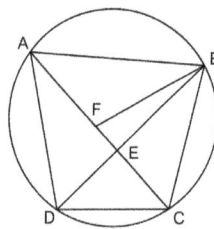

where F is such that $\angle ABF = \angle CBD$.

Problem 8.5 Suppose chords $\overline{AB}, \overline{CD}$ intersect at E, such that $AE : EB = 1 : 3$ and $CE : ED = 1 : 12$. Find the ratio of $AB : CD$.

Problem 8.6 (AHSME 1999 #21) A circle is circumscribed about a triangle with sides 20, 21, and 29, thus dividing the interior of the circle into four regions. Let $A, B,$ and C denote the areas of the non-triangular regions, with C being the largest. Compute $C - (A + B)$.

Problem 8.7 Let A, B, C, D be four points, arranged in clockwise order, on circle ω. Segments AC and BD intersect at P. Given that $AB = 3, BP = 4, PA = 5, PC = 6$, find the radius of circle ω.

Problem 8.8 Suppose we have a rectangle $ABCD$ with $AB = 8$, $BC = 12$. Inscribe a circle in the rectangle so that it touches sides $\overline{AB}, \overline{BC}, \overline{AD}$. Let M be the midpoint of $\overline{AB}$. Call $E \neq M$ the intersection of $\overline{MD}$ with the circle. Find DE.

Problem 8.9 Given right triangle ABC, construct semicircles on the three sides as shown.

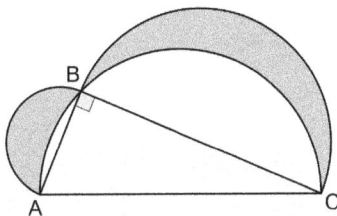

Given that $AB = 5, BC = 12$. Find the sum of the areas of the shaded regions.

Problem 8.10 In $\triangle ABC$, $AB = 37, AC = 58$. Use A as center and AB as radius, draw a circle to intersect $\overline{BC}$ at D where D is between B and C.

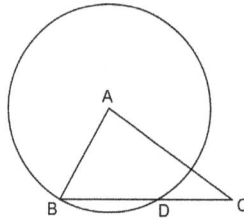

Given that the lengths of $\overline{BD}$ and $\overline{DC}$ are both integers, compute BC.

8.2 Quick Response Questions

Problem 8.11 Let $ABCD$ be a cyclic quadrilateral. Suppose $\angle DAB = 120°$ and $\angle BDC = 40°$, what is the measure of $\angle DBC$?

Problem 8.12 What is a better name for a cyclic parallelogram?

(A) Trapezoid
(B) Rhombus
(C) Rectangle
(D) None of the above

Problem 8.13 Let $ABCD$ be a quadrilateral with $\angle DAB = 116°$, $\angle ABC = 92°$, $\angle BCD = 60°$ and $\angle CDA = 92°$. Is it true that $ABCD$ is a cyclic quadrilateral?

Problem 8.14 Let $ABCD$ be a quadrilateral with $\angle DAB = 87°$, $\angle ABC = 57°$, $\angle BCD = 93°$ and $\angle CDA = 123°$. Is it true that $ABCD$ is a cyclic quadrilateral?

Problem 8.15 Suppose a trapezoid $ABCD$ with base $\overline{AD}$ is inscribed in a circle. If $AB = 4$, what is CD?

Problem 8.16 Let $\overline{PQ}$ be a chord with $PQ = 12$. Extend $\overline{PQ}$ to a point R such that $QR = 4$. Let T be such that $\overline{RT}$ is tangent to the circle. What is RT?

Problem 8.17 Let A, B, C and D be points on circle $\mathscr{C}$ such that AB and CD intersect at point P outside of $\mathscr{C}$. If $AB = 10$, $AP = 16$, and $CP = 15$, what is DP?

Problem 8.18 Let A, B, C and D be points on a circle $\mathscr{C}$ such that AB and CD intersect at point P inside of $\mathscr{C}$. If $AB = 10$, $AP = 4$, and $CP = 3$, what is DC?

Problem 8.19 Suppose O is the center of a unit circle. Let $ABCO$ be a rhombus with A, B, C on the circle. $[ABCO] = \frac{\sqrt{P}}{Q}$. What is $P + Q$?

Problem 8.20 Suppose equilateral triangle ABC is inscribed in a circle. Let P be on minor arc $\overset{\frown}{AB}$ such that $AP = 1$ and $BP = 3$ (distances of the line segments). Find the distance PC.

8.3 Practice Questions

Problem 8.21 Prove the Pythagorean theorem using Ptolemy's theorem.

Problem 8.22 Prove that if $ABCD$ is a cyclic quadrilateral, then $\angle A + \angle C = \angle B + \angle D = 180°$.

Problem 8.23 Suppose $ABCD$ is a quadrilateral. Let $\overline{AC}, \overline{BD}$ intersect at E. If $AE \cdot EC = BE \cdot ED$, show that $ABCD$ is cyclic.

Problem 8.24 Prove the Power of a Point formula for two chords $\overline{AB}$ and $\overline{CD}$ intersecting at E outside the circle.

Problem 8.25 Suppose diameter $\overline{CD}$ intersects chord $\overline{AB}$ at E, so that $AE = 4, EB = 9$. If the diameter of the circle is 15, Find CE and ED.

Problem 8.26 Suppose you have a circle with area 1. Inscribe a triangle in the circle. Let A, B, and C denote the area of the non-triangular regions, with C being the largest.

(a) Show that $C - (A + B)$ can get very close to 1.

(b) Find the minimum value of $C - (A + B)$. Hint: It is negative!

Problem 8.27 Suppose $ABCD$ is an isosceles trapezoid (so it is cyclic). Further suppose the center of this circle is in the interior of the trapezoid $ABCD$, and the radius is 25, with $AD = 40$ and $BC = 48$. Find the area of trapezoid $ABCD$.

Problem 8.28 Suppose we have a rectangle $ABCD$ with $AB = 8$, $BC = 12$. Inscribe a circle in the rectangle so that it touches sides $\overline{AB}, \overline{BC}, \overline{AD}$. Let M be the midpoint of $\overline{AB}$. Call $E \neq M$ the intersection of $\overline{MD}$ with the circle. Assume that CE is tangent to the circle at E (It is, as a challenge try to prove it!). Find the length of CE.

Problem 8.29 Inscribe equilateral $\triangle ABC$ in a circle of radius 1. Then construct semicircles on the three sides as shown.

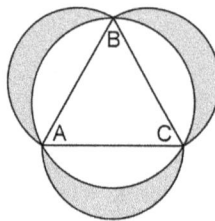

Find the sum of the areas of the shaded regions.

Problem 8.30 Let O be the circumcenter of $\triangle ABC$. Through A construct a line tangent to $\odot O$ and intersecting the extension of $\overline{BC}$ at D. From B and C construct lines perpendicular to $\overline{AD}$ with feet M and N respectively. Assume that $CN = 5, BC = 5, AD = 5\sqrt{6}$. Find the area of trapezoid $MBCN$.

9. Solid Geometry

Basics

- Points in three dimensions can be thought of as using x, y, z coordinates, written (x, y, z).
- We have analogues of the Pythagorean Theorem and Distance Formula in three dimensions: The distance from $(0, 0, 0)$ to (a, b, c) is $\sqrt{a^2 + b^2 + c^2}$.
- Given two distinct lines, there are three possibilities: (i) they are parallel, (ii) they intersect, or (iii) they are *skew*.
- Given two distinct planes, there are two possibilities: (i) they are parallel, or (ii) they intersect (and their intersection is a line).

Solids

- **Sphere**: The collection of points (in three dimensions) of equal distance (called the *radius* from a *center*).
  - A sphere with radius r has volume $\frac{4}{3}\pi r^3$.
  - A sphere with radius r has surface area $4\pi r^2$.
- **Cube**: The 3-D version of a square.
  - A cube with side length s has volume s^3.
  - A cube with side length s has surface area $6s^2$.
  - A cube has 8 vertices, 12 edges, and 6 faces. The 6 faces are all squares.
- **Rectangular Prism**: A "box".
  - A rectangular prism with sides l, w, h (often 'length', 'width', 'height') has volume lwh.

- A rectangular prism with sides l, w, h has surface area $2(lw + wh + lh)$.
- Like a cube, a rectangular prism has 8 vertices, 12 edges, and 6 faces. The 6 faces are all rectangles.
- **Cylinder**: A "can".
  - A cylinder with height h and radius r has volume $\pi r^2 h$.
  - A cylinder with height h and radius r has surface area $2\pi r^2 + 2\pi rh$.
- **Square Right Pyramid**: The standard "pyramid from Egypt" solid, with a square *base* and *apex* (or top point) that is centered above the square.
  - A square right pyramid with height h with square base of side length s has volume $\frac{1}{3}s^2 h$.
  - A square right pyramid with height h with square base of side length s has surface area $s^2 + 2sL$, where $L = \sqrt{h^2 + s^2/4}$ (L is called the *slant height*).
  - A square right pyramid has 5 vertices, 8 edges, and 5 faces. The 4 side faces are all triangles.
- **Right Cone**: The standard "ice cream cone" solid, with a circular *base* and *apex* that is centered above the circle.
  - A right with height h with square base of radius r has volume $\frac{1}{3}\pi r^2 h$.
  - A square right pyramid with height h with square base of side length s has surface area $\pi r^2 + \pi rL$, where $L = \sqrt{h^2 + r^2}$ (L is called the *lateral height*).
- **Tetrahedron**: A triangular pyramid. Alternatively a solid made up of four triangles.
  - As a tetrahedron is a triangular pyramid, the above formulas hold.
  - In particular, a *regular* tetrahedron is a tetrahedron made up of 4 equilateral triangles.

9.1 Example Questions

Problem 9.1 Find the volume of the largest sphere that can fit inside a cone of radius 1 and height $\sqrt{3}$.

Problem 9.2 Show that the surface area of a regular tetrahedron with side length a is $a^2\sqrt{3}$.

Problem 9.3 Show that the volume of a regular tetrahedron with side length a is $\dfrac{a^3\sqrt{2}}{12}$.

Problem 9.4 (2010 AMC 10A #20) Suppose a bored bee lives on a cube with side length 1. For "fun" he decides to visit every vertex of the cube, each exactly once, starting and ending at the same vertex. It will travel from one vertex to another using straight lines (either crawling or flying). Give an example of a path that uses the maximum distance and find this distance.

Problem 9.5 Four identical balls (spheres), each of radius 1 in, are glued to the ground so that their centers form the vertices of a square with side length 2 in. Suppose you rest a fifth identical ball on the four balls (so the fifth ball is a sphere externally tangent to the other spheres). How far does this ball rest off the ground?

Problem 9.6 Suppose you pick 4 vertices of a cube to form a tetrahedron.

(a) How many different (non-congruent) tetrahedra are possible? Are any of them regular tetrahedra?

(b) Find the volumes for each of the possibilities in (a) if the cube has volume 1.

Problem 9.7 Suppose you have a unit cube. Pick two opposite corners. In each corner, form a tetrahedron using the corner and the three adjacent vertices. Remove these two tetrahedra and call the resulting polyhedron $\mathscr{S}$.

(a) How many vertices, edges, and faces does the resulting polyhedron have? Describe the faces.

(b) Find the volume of $\mathscr{S}$.

Problem 9.8 Suppose you have a sphere of radius 1. That is the side length of the largest regular tetrahedron you can fit (inscribe) inside the sphere?

Problem 9.9 An obtuse triangle with dimensions 9, 10, and 17 is rotated about the smallest side so that it creates a three-dimensional solid shown below.

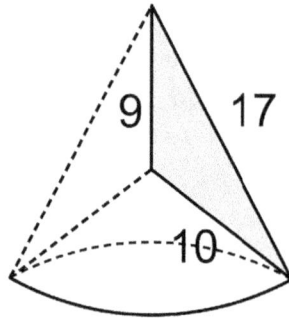

Determine the surface area of the solid. Use $\pi = 3.14$ and round your answer to the nearest tenth if necessary.

Problem 9.10 Suppose you start with a right cone and cut off the top of the cone with a plane parallel to the base. The resulting solid, called a *frustrum* has two circular "bases", say with radii R and r (with $R > r$), and height h. (Hence from the side the frustrum looks like a trapezoid with bases R, r and height H.)

(a) Show that the volume of a frustrum is $\dfrac{\pi H}{3}(R^2 + Rr + r^2)$.

(b) Suppose a sphere can be inscribed in a frustrum with base radii r, R such that the sphere is tangent to the two bases and the side. Find the radius of such a sphere in terms of r, R.

9.2 Quick Response Questions

Problem 9.11 How many total pairs of parallel edges does a cube contain?

Problem 9.12 Suppose we have a ball with radius 6. Suppose you cut the ball in half. What is the volume of the half-ball? Use $\pi = 3.14$ and round your answer to the nearest tenth if necessary.

Problem 9.13 Suppose we have a ball with radius 6. Suppose you cut the ball in half. What is the surface area of the half-ball? Use $\pi = 3.14$ and round your answer to the nearest tenth if necessary.

Problem 9.14 A cube is increased to form a new cube so that the surface area of the new cube is 64 times that of the original cube. By what factor is the volume of the cube increased?

Problem 9.15 A cube is increased to form a new cube so that the volume of the new cube is 64 times that of the original cube. By what factor is the surface area of the cube increased?

Problem 9.16 What is the volume of a square pyramid with base side length 6 and height 4? Round your answer to the nearest integer if necessary.

Problem 9.17 What is the surface area of a square pyramid with base side length 6 and height 4? Round your answer to the nearest integer if necessary.

Problem 9.18 In the $2 \times 2 \times 2$ cubic figure below, if the path is required to be along the surface of the cube, what is the length of the shortest path from point A to B. This length can be expressed in the form $\sqrt{K}$ for an integer K. What is K?

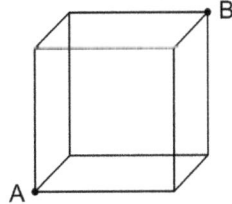

Problem 9.19 Suppose you have an ice cream cone with radius 2 inches and height 4 inches. The cone starts full of ice cream (but there is no ice cream outside the cone). After you've eaten some ice cream and some of the cone you are left with a cone with a radius and a height of 2 inches. What percent of the ice cream have you eaten? (For example, if your answer is 40%, input 40.)

Problem 9.20 A regular tetrahedron has volume $\dfrac{8}{3}$. What is the side length, rounded to the nearest tenth if necessary?

9.3   Practice Questions

Problem 9.21 Assume a sphere of radius $\dfrac{\sqrt{3}}{3}$ is put in a cone with radius 1 and height h. Suppose you now want to fit another sphere in the cone that is tangent to the base of the cone. Find the radius of the largest such sphere.

Problem 9.22 Suppose $S - ABC$ is a regular tetrahedron with apex S. Cut off the top half of the tetrahedron (that is, cut through the midpoints of $\overline{SA}, \overline{SB}, \overline{SC}$ and leave the bottom).

(a) How many vertices, edges, and faces does the resulting solid have?

(b) Find the volume and the surface area as ratios to the original volume and surface area.

Problem 9.23 Form a triangular pyramid with a base that is an equilateral triangle with side length 2. If the three sides of the pyramid are isosceles triangles with sides $4, 4, 2$, what is the volume of the pyramid?

Problem 9.24 Suppose a bored bee lives on a cube with side length 1. For fun he decides to visit every vertex of the cube, each exactly once, starting and ending at the same vertex. It will travel from one vertex to another using straight lines (either crawling or flying).

(a) What is the distance of the shortest such path?

(b) Note in part (a) the shortest path was on the surface of the cube (that is, did not travel inside the cube). What is the longest path if the bee is not allowed to travel inside the cube?

Problem 9.25 Four identical balls (spheres), each of radius 1in, are glued to the ground so that their centers form the vertices of a square with side length 2in. Suppose you rest a fifth ball that rests 1in off the ground. Find the radius of the fifth ball.

Problem 9.26 There is one regular tetrahedron that can be formed by choosing 4 vertices from a unit cube. What is the surface area of this tetrahedron?

Problem 9.27 As in class, form a solid by removing opposite tetrahedra in a unit cube. (Here each tetrahderon is formed by a vertex and the three adjacent vertices in a cube.) Suppose $\mathscr{S}$ is resting on one of the faces (ignore whether the polyhedron would actually balance or not). What are different possible heights of $\mathscr{S}$?

Problem 9.28 Inscribing Spheres and Cubes

(a) Find the volume of the largest sphere that fits in a cube of volume 1. (That is, inscribe a sphere inside the cube.)

(b) Find the volume of the smallest sphere that holds a cube of volume 1. (That is, circumscribe a sphere outside the cube.)

Problem 9.29 An obtuse triangle with dimensions 9, 10, and 17 is rotated about the smallest side so that it creates a three-dimensional solid shown below.

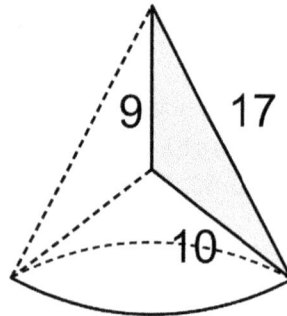

Determine the volume of the solid.

Problem 9.30 Recall the in class questions which discussed frustrums and when it was possible to inscribe a sphere in it. A sphere is inscribed in a frustrum with bases of radii 1 and 2. The intersection of this sphere with the side of the frustrum is a circle. Find the radius of this circle.

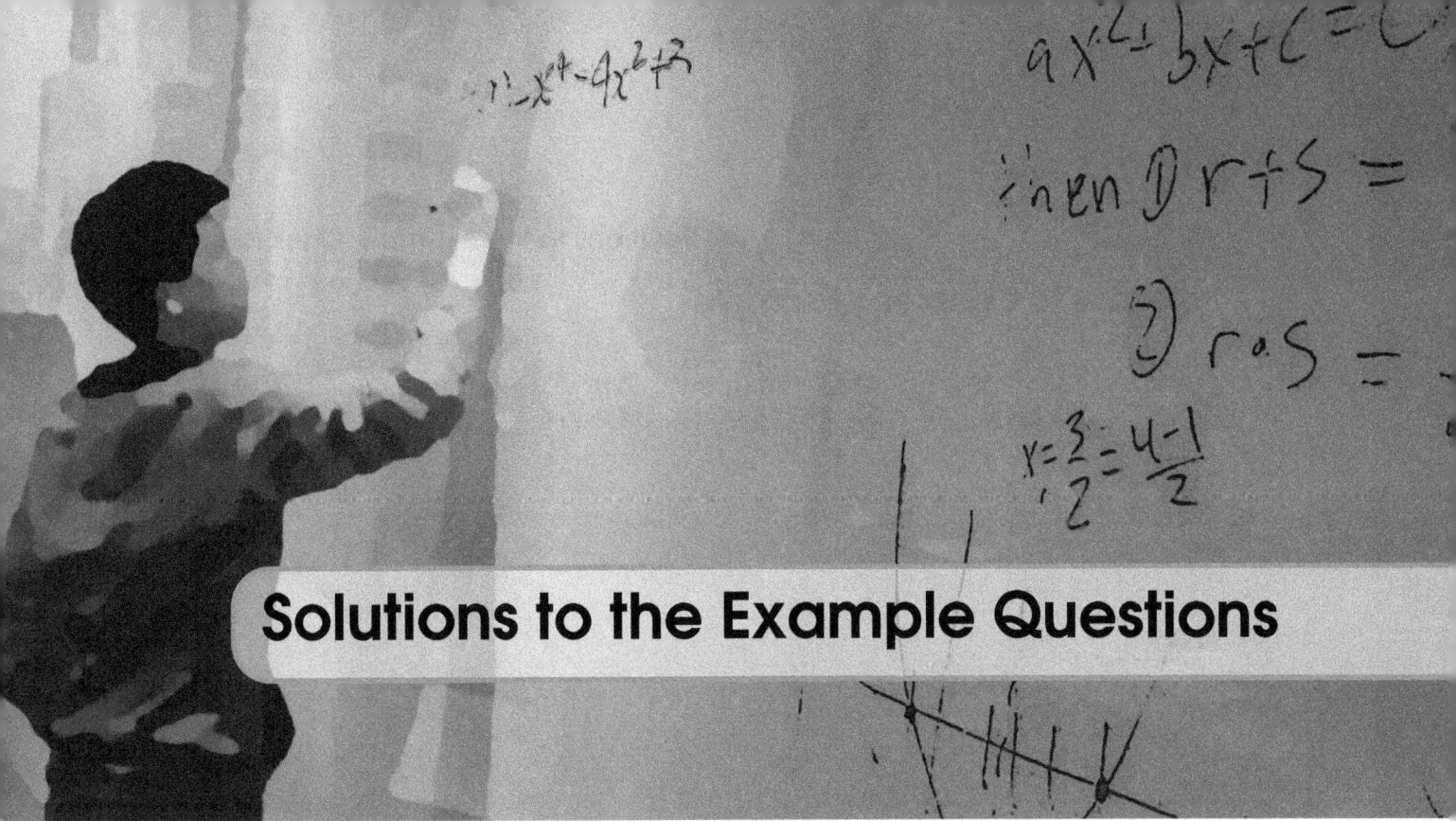

Solutions to the Example Questions

In the sections below you will find solutions to all of the Example Questions contained in this book.

Quick Response and Practice questions are meant to be used for homework, so their answers and solutions are not included. Teachers or math coaches may contact Areteem at info@areteem.org for answer keys and options for purchasing a Teachers' Edition of the course.

1   Solutions to Chapter 1 Examples

Problem 1.1 Suppose that *ABCD* is a square, and that *CDP* is an equilateral triangle, with *P outside* the square. What is the size of angle *PAD*?

Answer

$15°$.

Solution

We have $\angle ADP = \angle ADC + \angle CDP = 90 + 60 = 150°$. Since $\triangle ADP$ is isosceles $\angle PAD = (180 - 150)/2 = 15°$.

Problem 1.2 Given square *ABCD*, let *P* and *Q* be the points outside the square that make triangles *CDP* and *BCQ* equilateral. Segments *AQ* and *BP* intersect at *T*. Find angle *ATP*.

Answer

$90°$.

Solution

$\triangle ABQ \cong \triangle BCP$ (why?), therefore $\angle BAT = \angle TBC$ (equals $15°$ by a previous problem). Consider $\triangle ABT$. Since $\angle BAT = \angle TBC$, $\angle BAT + \angle ABT = 90°$ and thus $\angle ATB = 90°$. This implies $\angle ATP = 90°$ as well.

Problem 1.3 Squares *OPAL* and *KEPT* are attached to the outside of equilateral triangle *PEA*. Draw segment *TO*, then find the size of angle *TOP*.

Answer

$30°$.

Solution

We have $\angle TPO = 360 - 90 - 60 - 90 = 120°$. As $\triangle TOP$ is isosceles $\angle TOP = (180 - 120)/2 = 30°$.

Problem 1.4 Mark *P* inside square *ABCD*, so that triangle *ABP* is equilateral. Let

---

Q be the intersection of BP with diagonal AC. Triangle CPQ looks isosceles. Is this actually true?

Answer

Yes.

Solution

Using isosceles triangles, it is easy to see that $\angle BPC = \angle BCP = 75°$. Since $\angle BCA = 45°$, we get $\angle PCQ = 75° - 45° = 30°$. Hence $\angle PQC = 75° = \angle CPQ$. So $\triangle CPQ$ is a $30°$-$75°$-$75°$ triangle.

Problem 1.5 Let triangle ABC be equilateral triangle with side length 16. Let D be on side $\overline{AB}$ and E be on side $\overline{AC}$ such that $\overline{DE} \| \overline{BC}$. Assume triangle ADE and trapezoid $DECB$ have the same perimeter. What is the length of $\overline{AD}$?

Answer

12.

Solution

Let $x = AD$. Note that $\triangle ADE$ is equilateral (why?). Thus the perimeter of $\triangle ADE = 3x$ and the perimeter of trapezoid $DECB$ is $x + 2(16 - x) + 16 = 48 - x$. Since these two perimeters are equal, $3x = 48 - x$ so $x = 12$.

Problem 1.6 Let E be a point inside unit square $ABCD$ such that CDE is an equilateral triangle. Find the area of triangle AEC.

Answer

$(\sqrt{3} - 1)/4$.

Solution

Note that $[AEC] = [AED] + [CDE] - [ADC]$. Further, (why?) $[AED] = \frac{1}{2} \cdot 1 \cdot \frac{1}{2}, CDE = \frac{1}{2} \cdot 1 \cdot \frac{\sqrt{3}}{2} = \frac{\sqrt{3}}{4}, [ADC] = \frac{1}{2} \cdot 1 \cdot 1$. Hence, $[AEC] = \frac{1}{4} + \frac{\sqrt{3}}{4} - \frac{1}{2} = \frac{\sqrt{3} - 1}{4}$.

Problem 1.7 A triangle has a 60-degree angle and a 45-degree angle, and the side opposite the 45-degree angle has length 12. How long is the side opposite the 60-degree angle?

Answer

$6\sqrt{6}$.

Solution

It is easy using the Law of Sines. Solution without the Law of Sines is as follows. The remaining angle is $75°$. Draw the altitude from the vertex of $75°$, and this altitude has length $6\sqrt{3}$ based on the 30-60-90 triangle, and then the required side length is $6\sqrt{6}$ based on the 45-45-90 triangle.

Problem 1.8 Let ABC be a triangle with $AB = AC$ and $\angle BAC = 20°$, and let P be a point on side AB such that $AP = BC$. Construct point D such that triangle ACD is equilateral, as shown in the diagram below. Show that triangle DCP is isosceles.

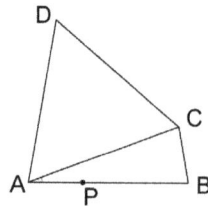

Solution

Connect $\overline{PD}$. Then $\triangle ABC \cong \triangle DAP$ by SAS congruency (double check this!). Thus $DP = AC = DC$ and thus $\triangle DCP$ is isosceles.

Problem 1.9 In triangle ABC, $AB = AC$, and BD is the altitude on AC. Given that $BD = \sqrt{3}$, and $\angle DBC = 60°$, find the area of $\triangle ABC$.

Answer

$\sqrt{3}$.

Solution

First note that since $\angle DBC = 60°$, we must have the altitude BD intersects the extension of AC, as in the diagram below.

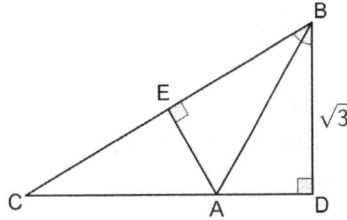

Since $\triangle BCD$ is a 30-60-90 triangle, we have $BC = 2\sqrt{3}$. Letting E be the midpoint of BC (so also AE is the altitude on BC) we have $\triangle AEC \cong \triangle AEB \cong \triangle ADB$ and all are 30-60-90 triangles themselves. Hence $AE = 1$ so the area of ABC is $\frac{1}{2} \cdot 2\sqrt{3} \cdot 1 = \sqrt{3}$ (using BC as the base).

Problem 1.10 In a right triangle $\triangle ABC$ suppose $\angle B = 90°$ and $\angle C = 30°$. Suppose point D is on $\overline{BC}$ with $\angle ADB = 45°$ and $DC = 10$. Find the length of AB.

Answer

$10/(\sqrt{3}-1)$.

Solution

Let $x = AB$. Then as $\triangle ABD$ is a 45-45-90 triangle, so $DB = x$ as well. As $\triangle ABC$ is a 30-60-90 triangle, $BC = \sqrt{3}AB$ so $x + 10 = x\sqrt{3}$. Hence $x = \dfrac{10}{\sqrt{3}-1}$.

2   Solutions to Chapter 2 Examples

Problem 2.1 Complete the following table about polygons with n sides: name, sum of interior angles, sum of exterior angles, and measure of each angle in case of regular polygon. All angles are in degrees. Justify your answers. Keep the chart for your own reference.

n	Name	Int. Angle Sum	Ext. Angle Sum	Each Angle (if regular)
3	Triangle			
4				
5				
6				
7	Heptagon			
8				
9	Nonagon			
10				
12	Dodecagon			
20	Icosagon			

Solution

To justify the sum of interior angles of a triangle: In $\triangle ABC$, draw line through A that's parallel to $\overline{BC}$, then use the property of alternate interior angles. For polygons of more sides: cut them into triangles, and get the formula $(n-2)180°$.

For sum of exterior angles: Pretend the polygon's sides are streets, you are walking along the streets. Each time you turn a corner, you turn an exterior angle; when you return to the starting point, you turned a total of $360°$. You can also use the formula to calculate it. This gives the table below:

n	Name	Int. Angle Sum	Ext. Angle Sum	Each Angle (if regular)
3	Triangle	180°	360°	60°
4	Quadrilateral	360°	360°	90°
5	Pentagon	540°	360°	108°
6	Hexagon	720°	360°	120°
7	Heptagon	900°	360°	900/7°
8	Octagon	1080°	360°	135°
9	Nonagon	1260°	360°	140°
10	Decagon	1440°	360°	144°
12	Dodecagon	1800°	360°	150°
20	Icosagon	3240°	360°	162°

Problem 2.2 Use 6 equilateral triangles to form a hexagon $ABCDEF$.

(a) Show hexagon $ABCDEF$ is regular. Justify your answer.

Solution

It is clear that each side has the same length. Each angle in the hexagon is made up of two 60° angles (from the equilateral triangle), hence each angle is 120°.

(b) Calculate the angle AED.

Answer

90°

Solution

Note that $\triangle AFE$ is isosceles with $\angle AFE = 120°$. Thus, $\angle FEA = (180 - 120)/2 = 30°$ and $\angle AED = \angle FED - \angle FEA = 120 - 30 = 90°$.

Problem 2.3 Four non-overlapping regular plane polygons all have sides of length 1. The polygons meet at a point A in such a way that the sum of the four interior angles at A is $360°$. Among the four polygons, two are squares and one is a triangle. What is the last polygon?

Answer

Hexagon

Solution

The other polygon's interior angle is $120°$, so it is a hexagon.

Problem 2.4 Find the area of the largest equilateral triangle that fits in a regular hexagon of area 50.

Answer

25

Solution

If the hexagon is $ABCDEF$, the largest equilateral triangle is ACE (or equivalently BDF). If O is the center of the hexagon, note that $\triangle ABC \cong \triangle ACO$ (why?). Similar results hold for the rest of the hexagon, and from there it is not hard to show that $[ACE] = [ABCDEF]/2 = 25$.

Problem 2.5 In equiangular octagon $ABCDEFGH$, $AB = CD = EF = GH = 6\sqrt{2}$ and $BC = DE = FG = HA$. Given the area of the octagon is 184, compute the length of side BC.

Answer

4

Solution 1

Let $x = BC$, connect $\overline{AD}, \overline{EH}, \overline{BG}, \overline{CF}$. The sides $\overline{AB}, \overline{CD}, \overline{EF}, \overline{GH}$ are hypotenuses of four isosceles right triangles, whose legs are equal to 6. Thus the area of the whole octagon is

$$x^2 + 4 \cdot 6 \cdot x + 4 \cdot \frac{1}{2} \cdot 6 \cdot 6 = 184,$$

so

$$x^2 + 24x - 112 = 0.$$

We want the positive root: $x = 4$.

Solution 2

A better solution involves extending the sides $\overline{BC}, \overline{DE}, \overline{FG}, \overline{HA}$ to both directions to make a big square. The added corners of this square are isosceles right triangles with hypotenuses $6\sqrt{2}$, so their legs are all 6. Let $x = BC$, the big square's side length is $x + 12$. The added corners have a total area of two squares of side 6, so the added area is 72. Thus $(x + 12)^2 = 184 + 72 = 256$, thus $x + 12 = 16$, and $x = 4$.

Problem 2.6 Answer the following

(a) Given a 60-120 isosceles trapezoid, prove that the sum of the length of the top and one of the side is equal to the length of the base.

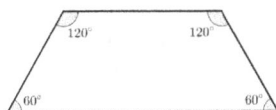

Solution

Divide the trapezoid as follows:

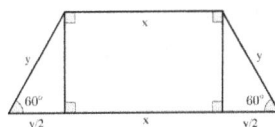

Let x denote the length of the top of the trapezoid and y denote the lengths of each side. If the trapezoid is divided into a rectangle and two 30-60-90 Triangles as shown, it is easy to see the length of the base of the trapezoid is equal to $y/2 + x + y/2 = x + y$.

(b) Let triangle ABC be a equilateral triangle with length equals to 1. Let point P be the center of the triangle ABC, D be point on $\overline{BC}$, E be point on $\overline{CA}$, and F be point on $\overline{AB}$, so that $\overline{PD} \| \overline{AB}$, $\overline{PE} \| \overline{BC}$, and $\overline{PF} \| \overline{AC}$. Find the value of $PD + PE + PF$.

Answer

1

Solution

Note that $PDCE, PFBD, PEAF$ are all isosceles trapezoids. By the isosceles trapezoid problem, $PD + PE = CD$. Further, $PF = DB$, so $PD + PE + PF = CD + DB = CB = 1$.

Problem 2.7 Show that a regular dodecagon (12-sided polygon) can be cut into pieces that are all regular polygons, which need not all have the same number of sides.

Solution

A regular hexagon at the center, surrounded by 6 squares and 6 equilateral triangles, all having the same side length.

Problem 2.8 What is the side length of the largest equilateral triangle that can fit inside a 2-by-2 square?

Answer

$2(\sqrt{6} - \sqrt{2})$.

Solution

In square $ABCD$, let the largest equilateral triangle be $\triangle AEF$, where E is on BC and F is on CD. Let $x = EC = CF$, then $BE = DF = 2 - x$. So $2^2 + (2 - x)^2 = 2x^2$, solving for x, we get $x = 2(\sqrt{3} - 1)$, and the side length of $\triangle AEF$ is $2(\sqrt{6} - \sqrt{2})$.

Problem 2.9 Let $ABCD$ be a unit square and let P, Q be on sides $\overline{AD}, \overline{AB}$ respectively

---

such that $\triangle APQ$ has perimeter 2. Rotate $\triangle PDC$ $90°$ about C. Call the point P is rotated to P'. Prove that $\triangle PQC$ is congruent to $\triangle P'QC$.

Solution

Since $ABCD$ is a unit square, it is not hard to show that PQ has the same length as QP'. Then use SSS to show the triangles are congruent.

Problem 2.10 In convex quadrilateral $ABCD$, $\angle A = 60°$, $\angle C = 30°$, and $AB = AD$. Show that $AC^2 = BC^2 + CD^2$.

Solution

First recognize that $\triangle ABD$ is equilateral. Construct equilateral triangle ACC' where B is in the interior of $\triangle ACC'$ as in the diagram below.

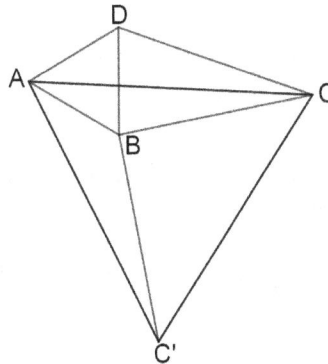

Since $AD = AB$, $\angle DAC = \angle BAC' = 60° - \angle CAB$, and $AC = AC'$, we have $\triangle ADC \cong \triangle ABC'$. Thus $\angle ABC' = \angle ADC$. Since $\angle ADC + 60° + \angle ABC + 30° = 360°$, and $\angle ABC' + \angle ABC + \angle CBC' = 360°$, it follows that $\angle CBC' = 90°$. So by Pythagorean Theorem, $CC'^2 = BC^2 + BC'^2$, thus $AC^2 = BC^2 + CD^2$.

3   Solutions to Chapter 3 Examples

Problem 3.1 (Not analytic geometry) Prove that in a right triangle $\triangle ABC$ (with $\angle B = 90°$), that the perpendicular bisectors of $\overline{AB}$ and $\overline{BC}$ meet at the midpoint of AC.

Solution

Let F be the midpoints of $\overline{AC}$. Let D, E on $\overline{AB}, \overline{BC}$ such that $BEFD$ is a rectangle. Hence $BE = DF$ and $EF = BD$. Further, using parallel lines, $\angle FAD = \angle CFE$, so $\triangle FAD \cong \triangle CFE$ using AAS. Hence $EC = DF = BE$ and $AD = FE = DB$ so D, E are midpoints and hence $\overline{DF}$ and $\overline{EF}$ are the perpendicular bisectors.

Problem 3.2 Prove that the equation for the line going through points A and B is given by

$$\frac{x - x_A}{x_A - x_B} = \frac{y - y_A}{y_A - y_B}.$$

Solution

Note that a point $P = (x, y)$ is on the same line as A and B if and only if the slope from P to A and A to B are the same. That is

$$\frac{y - y_A}{x - x_A} = \frac{y_A - y_B}{x_A - x_B} \text{ or (after rearranging) } \frac{x - x_A}{x_A - x_B} = \frac{y - y_A}{y_A - y_B}$$

as needed.

Problem 3.3 Suppose a triangle has vertices $(3, 4), (4, 7), (7, 6)$. Find the area of the triangle. Hint: One possible method is to find the altitude from $(4, 7)$.

Answer

5.

Solution

Calculate the line between $(3, 4), (7, 6)$ as $-x + 2y = 5$ with slope $\frac{1}{2}$. Hence a perpendicular line has slope -2, and the one containing $(4, 7)$ is $2x + y = 15$. These two lines intersect at $(5, 5)$, so using the distance formula we can calculate the altitude has length $\sqrt{5}$. As the distance between $(3, 4), (7, 6)$ is $2\sqrt{5}$, the area of the triangle is 5.

Problem 3.4 Suppose $\triangle ABC$ is an equilateral triangle with with $A = (0,2)$ and $B = (\sqrt{3},1)$. Find all possible coordinates for C.

Answer

$(0,0), (\sqrt{3},3)$.

Solution

Note the distance from A to B is 2 and the slope from A to B is $-\sqrt{3}$. This slope is an angle of $-30°$ from the x-axis. As $\triangle ABC$ is equilateral, all possible choices for C must be distance 2 from A. Since an equilateral triangle has angles of $60°$, the two possible points must be form a $-30+60 = 30°$ or $-30-60 = -90°$ angle with A and the x-axis. These leads to the possibilities of $(0,0)$ or $(\sqrt{3},3)$.

Problem 3.5 Suppose you have a line $\ell : Ax + By + C = 0$ and a point P.

(a) If $A = 0$ find the (shortest) distance from P to ℓ.

Answer

$$\left| \frac{-C}{B} - y_P \right| = \left| \frac{B \cdot y_P + C}{B} \right|.$$

Solution

If A is zero, ℓ is a horizontal line $y = -C/B$. Therefore the shortest path from P to ℓ is a vertical line from P to ℓ, having distance as above.

(b) If $B = 0$ find the (shortest) distance from P to ℓ.

Answer

$$\left| \frac{-C}{A} - x_P \right| = \left| \frac{A \cdot x_P + C}{A} \right|.$$

Solution

Similar to part (a) except now ℓ is a vertical line, so the shortest path is a horizonal line.

(c) The general equation for the (shortest) distance d from P to ℓ is

$$d = \frac{|A \cdot x_P + B \cdot y_P + C|}{\sqrt{A^2 + B^2}}.$$

Verify this equation for the line $3x + 4y + 6 = 0$ and $P = (8, -5)$. That is, calculate the distance WITHOUT using the formula, and check your work with the formula.

> **Solution**

Note in the previous parts, the shortest path is always on a line perpendicular to ℓ. The line perpendicular to the line $3x + 4y + 6 = 0$ that contains the point $(8, -5)$ can be calculated to be $-4x + 3y + 47 = 0$, which intersects the original line at the point $(34/5, -33/5)$. This point has distance 2 to P, which matches the formula given.

Problem 3.6 Prove that quadrilateral $ABCD$ is a parallelogram if and only if

$$\begin{cases} x_A + x_C & = & x_B + x_D, \\ y_A + y_C & = & y_B + y_D. \end{cases}$$

> **Solution**

Recall that a quadrilateral is a parallelogram if and only if its diagonals bisect each other (prove this using similar triangles). The diagonals bisect each other if and only if the midpoint of $\overline{AC}$ is equal to the midpoint of BD. It is routine to check this is equivalent to the above equations.

Problem 3.7 Circles

(a) Prove (not necessarily using analytic geometry) that if a circle has diameter $\overline{AB}$, then a point P lies on the circle if and only if $\angle APB = 90°$.

> **Solution**

Consider the diagram below:

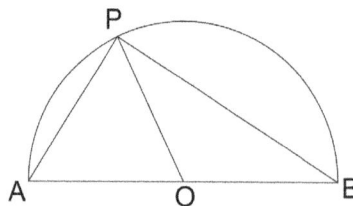

Note that OA, OB, OP are all radii, so $\triangle OAP, \triangle OPB$ are both isosceles. Since $\angle AOP + \angle BOP = 180°$, $2\angle OPA + 2\angle OPB = 180°$ so $\angle APB = \angle OPA + \angle OPB = 90°$ as needed. Further, for any point Q on the ray $\overrightarrow{OP}$ is Q is inside the circle then $\angle AQB > 90°$ and if Q is outside the circle $\angle AQB < 90°$.

(b) Prove that if $\overline{AB}$ is the diameter of a circle, then the circle has equation $(x - x_A)(x - x_B) + (y - y_A)(y - y_B) = 0$.

Solution

Using the part above, if $P = (x, y)$ is a point on the circle, we need the slopes of $\overline{AP}$ and $\overline{PB}$ to be multiply to -1, that is:

$$\frac{y - y_A}{x - x_A} \cdot \frac{y - y_B}{x - x_B} = -1 \text{ or clearing fractions and rearranging } (x - x_A)(x - x_B) + (y - y_A)(y - y_B) = 0,$$

as needed.

Problem 3.8 One definition of a parabola is the collection of point that are equidistant from a point F (called the focus) and a line d (called the directrix). Verify that the parabola $y = \frac{1}{4}x^2$ consists of the points equidistant from $F : (0, 1)$ and $d : y = -1$.

Solution

Let P be an arbitrary point on the parabola, say $P = (x, x^2/4)$. Hence, the distance from P to the directrix is $x^2/4 + 1$. The distance from P to the focus is

$$\sqrt{(x - 0)^2 + (x^2/4 - 1)^2} = \sqrt{x^4/16 + x^2/2 + 1} = \sqrt{(x^2/4 + 1)^2} = (x^2/4 + 1)$$

as needed.

Problem 3.9 Find the shortest path starting and ending at the origin that goes around the circle $(x - 4)^2 + y^2 = 8$.

Answer

$4\sqrt{2}(1 + 3\pi/2)$.

Solution

Let A be the origin, and B be the center of the circle. Note the circle has radius $2\sqrt{2}$ and $AB = 4$. Hence the tangent lines to the circle will form angles of $45°$ with AB (as a right triangle with hypoteneuse 4 and side $2\sqrt{2}$ must be a 45-45-90 triangle). It is easy to see that the tangent lines touch the circle at $(2, \pm 2)$. The shortest path thus starts at the origin, follows one tangent line to the circle, then loops around the circle to the other tangent line and back to the original. As we are travelling $270/360 = 3/4$ of the circle, the total distance is thus

$$2\sqrt{2} + \frac{3}{4} \cdot 2\pi \cdot 2\sqrt{2} + 2\sqrt{2}$$

as needed.

Problem 3.10 A parabola has equation $y = x^2 + bx + c$ and the line $y = 5$ intersects the parabola at $x = -1, 3$.

(a) Find b and c.

Answer

$b = -2, c = 2$.

Solution

We have $5 = 1 - b + c$ and $5 = 9 + 3b + c$. Solving for b, c gives $b = -2, c = 2$.

(b) Find the vertex of the parabola.

Answer

$(1, 1)$.

Solution

Using symmetry, we see the x-coordinate of the vertex must be 1, so plugging into the equation we have the y-coordinate is also one.

(c) Find the focus and directrix of the parabola.

Answer

Focus: $\left(1, \dfrac{5}{4}\right)$, Directrix: $y = \dfrac{3}{4}$.

Solution

We know the vertex is $(1, 1)$ so the x-coordinate of the focus must be 1. Further, we know that the focus and directrix must be $(1, 1+k)$, $y = 1-k$ for some constant k. Since $(0, 2)$ is on the parabola, we know that (distance to focus and directrix is the same). Hence $\sqrt{1 + (1-k)^2} = 1 + k$ or after squaring both sides, $k^2 - 2k + 2 = k^2 + 2k + 1$ so $4k = 1$ and hence $k = 1/4$.

4   Solutions to Chapter 4 Examples

Problem 4.1 Using only the basics about parallel lines and congruent/similar triangles and the fact that the area of a rectangle is bh, prove the following. (Note: once a fact is proven below, you can use it in later parts.)

(a) The area of a parallelogram is bh.

Solution

As in the diagram below, cutting a triangle at one end of the parallelogram and moving it to the other results in a $b \times h$ rectangle.

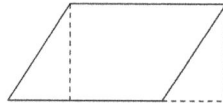

(b) The area of a triangle is $\frac{1}{2}bh$ (prove this two ways!).

Solution

Method 1: A right triangle is half a rectangle, so has area $\frac{1}{2}bh$. Then any triangle can be split into two right triangles (by dropping an altitude to the longest side).
Method 2: Two copies of any triangle can be combined to form a parallelogram with base b and height h.

(c) The area of a trapezoid is $\frac{b_1+b_2}{2}h$.

Solution

Draw a diagonal of the trapezoid. This breaks the trapezoid into two triangles, each with height h and having bases b_1 and b_2.

(d) The area of a trapezoid is also mh where m is the *median* of the trapezoid, which connects the midpoints of the two non-parallel sides.

Solution

As in the diagram below, cutting triangles below the median and rotating them above the median results in a $m \times h$ rectangle.

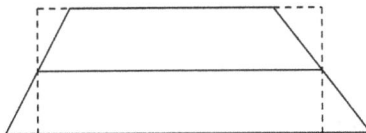

Problem 4.2 What is the ratio of areas between two similar triangles? Prove your result!

Solution

The square of the ratio between the corresponding sides.
Note that heights of similar triangles are also in ratio of the corresponding sides (as the altitude divides the similar triangles into smaller similar triangles).

Problem 4.3 Prove the Pythagorean Theorem using areas.

Solution

There are literally hundreds of different proofs. The simplest: Let $\triangle ABC$ be a right triangle where $\angle C$ is a right angle. Label the sides with a, b, c in the standard way. Draw the altitude $\overline{AD}$ on the hypotenuse. Then the three triangles, $\triangle ABC, \triangle ADC$ and $\triangle BDC$ are all similar. Their side ratio is $a : b : c$ based on the hypotenuses. Thus their area ratio is $a^2 : b^2 : c^2$. Since the two smaller triangle's areas add up to that of the big triangle, we have $a^2 + b^2 = c^2$.

Problem 4.4 In $\triangle ABC$, $AB > AC > BC$, $\overline{CD}, \overline{BE}, \overline{AF}$ are altitudes on $\overline{AB}, \overline{AC}, \overline{BC}$, respectively. Show that $CD < BE < AF$.

Solution

The area of $\triangle ABC$ can be calculated in three ways: $[ABC] = \dfrac{1}{2}AB \cdot CD = \dfrac{1}{2}AC \cdot BE = \dfrac{1}{2}BC \cdot AF$, so $CD/BE = AC/AB$ and $AF/BE = AC/BC$, thus gives the proof.

Problem 4.5 In triangle ABC, $AC = 10$, $BC = 24$, $AB = 26$. What is the altitude on $\overline{AB}$?

Answer

$\dfrac{120}{13}$.

Solution

Note $\triangle ABC$ is a right triangle. Therefore the area is $10 \cdot 24/2 = 120$. Since $AB = 26$ the area is also $26h/2 = 13h$ where h is the altitude on $\overline{AB}$, we have $h = 120/13$.

Problem 4.6 Let $ABCD$ be a parallelogram, and E, H, F, G be points on sides $\overline{AB}, \overline{BC}, \overline{CD}, \overline{DA}$ respectively, and $\overline{EF} \| \overline{BC}$ and $\overline{GH} \| \overline{AB}$. Let P be the intersection of $\overline{EF}$ and $\overline{GH}$. If $[GPFD] = 10, [PHCF] = 8, [EBHP] = 16$, find $[ABCD]$.

Answer

54.

Solution

Note $\overline{EF}, \overline{GH}$ divide $ABCD$ into four parallelograms, whose areas are proportional. That is $[AEPG]/[GPFD] = [EBHP]/[PHCF]$ so $[AEPG] = 10 \cdot 16/8 = 20$. Hence the total area of $ABCD$ is 54.

Problem 4.7 Suppose you only know that the centroid exists. Prove (using areas!) that the centroid divides each median in a ratio of $1 : 2$.

Solution

Consider the following diagram with the medians intersecting at G:

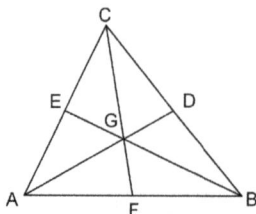

Note first that $[ACF] = [BCF]$ and $[AGF] = [BGF]$ (F is a midpoint and they share respective heights). Hence $[ACG] = [ACF] - [AGF] = [BCF] - [BGF] = [BCG]$. An identical argument gives $[ACG] = [ABG]$. Hence, $[AFG] = [ABG]/2 = [ACG]/2$ so in particular $[AFG] = [AFC]/3$. These two triangles share a height from A, so their bases FG, FC must be in ratio $1 : 3$ and hence $FG : GC = 1 : 2$. The other medians are handled identically.

Problem 4.8 Let $ABCD$ be a parallelogram, with midpoints E, F, G, H (say on $\overline{AB}, \overline{BC}, \overline{CD}, \overline{DA}$). Let I, J be the midpoints of $\overline{EF}, \overline{GH}$. Find the area of $\triangle JIG$ as a fraction of the area of $ABCD$.

Answer

$1/8$.

Solution

We will use the facts (which you should be able to prove!) $\overline{FH} \| \overline{DC}$ and $\overline{EF} \| \overline{GH}$. We have $[JIG] = [HIG]/2$ (same height, J midpoint), $[HIG] = [HFG]$ ($\overline{EF} \| \overline{GH}$), $[HFG] = [DCFH]/2$ ($\overline{FH} \| \overline{DC}$), and finally $[DCFH] = [ABCD]/2$ (F, H midpoints). Combining these give $[JIG] = [ABCD]/8$.

Problem 4.9 Prove that if a triangle has side lengths a, b, c, inradius r, and circumradius R we have $2Rr = \dfrac{abc}{a+b+c}$.

Solution

We have that the area of the triangle is $abc/4R$ and sr where s is the semiperimeter, so $abc = 4srR$. Rearranging gives the desired result.

Problem 4.10 Let $ABCD$ be a parallelogram as in the diagram, with E the midpoint of $\overline{BC}$.

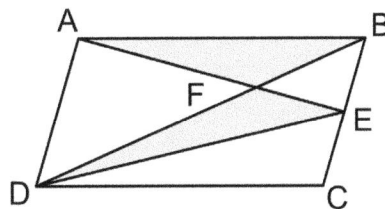

(a) Compare the shaded regions $\triangle ABF$ and $\triangle DEF$, which one has the larger area?

Answer

The have the same area.

Solution

Note $[ABD] = [DEF]$ as they have the same height and base. Hence, after removing the shared $\triangle AFD$ we see the two shaded triangles have the same area.

(b) Find the area of the shaded regions $\triangle ABF$ and $\triangle DEF$ in terms of the entire area of the parallelogram.

Answer

$[ABCD]/3$.

Solution

First note (since E is a midpoint) that $[DEC] = [BED] = [ABCD]/4$. Since $ABCD$ is a parallelogram and $BE = AD/2$, $\triangle AFD \sim \triangle EFB$, with ratio of side lengths $2:1$. Using this information, we have $[DFE] = 2[BEF]$. Therefore, $[DFE] = [ABF] = [ABCD]/6$. Hence the sum of the shaded regions is $[ABCD]/3$.

5 Solutions to Chapter 5 Examples

Problem 5.1 Prove that the perpendicular bisectors are concurrent (and therefore, the circumcenter is well-defined).

Solution

In triangle $\triangle ABC$, let D, E, F be the midpoints of $\overline{AC}, \overline{BC}, \overline{AB}$ respectively and let G be the intersection of the perpendicular bisectors of $\overline{AC}, \overline{BC}$. Using SAS we have $\triangle BEG \cong \triangle CEG$ and $\triangle CDG \cong \triangle ADG$. Note further that now using SSS we have $\triangle AFG \cong \triangle BFG$. Hence as $\angle AFG, \angle BFG$ are adjacent and equal, $\overline{FG}$ is actually the perpendicular bisector as needed.

Problem 5.2 Prove that the altitudes of a triangle are concurrent (and therefore, the orthocenter is well-defined). Use the previous result and the following hint: Given triangle $\triangle ABC$ construct lines through each vertex parallel to the opposite side to form $\triangle PQR$.

Solution

Note that vertices A, B, C are in fact the midpoints of the sides of $\triangle PQR$ (using parallelograms). Hence, the altitudes of $\triangle ABC$ are the same as the perpendicular bisectors of $\triangle PQR$ and hence are concurrent.

Problem 5.3 Prove the Angle Bisector Theorem using the following hint: In $\triangle ABC$, let E be on ray $\overrightarrow{AD}$ such that $\overline{CE} \parallel \overline{AB}$.

Solution

Note $\angle CAD = \angle CED$ (alternate interior), so $\triangle ACE$ is isosceles and $AC = CE$. Further, $\angle ABD = \angle BCE$ (alternate interior), and therefore using AAA we have $\triangle ABD \sim \triangle ECD$. Hence (as $AC = CE$), we have $\dfrac{AB}{BD} = \dfrac{CE}{DC} = \dfrac{AC}{DC}$.

Problem 5.4 Suppose a circle is drawn outside a triangle, so that the circle passes through each vertex of the triangle (the circle is *circumscribed* on the triangle). Find, with proof, the center of this circle. Can you guess another name for this circle?

Answer

Circumcenter and circumcircle.

Solution

We proceed very similarly to the proof that the perpendicular bisectors are concurrent. In triangle $\triangle ABC$, let D, E, F be the midpoints of $\overline{AC}, \overline{BC}, \overline{AB}$ respectively and let G be the intersection of the perpendicular bisectors of $\overline{AC}, \overline{BC}, \overline{AB}$. Note using SAS we have $\triangle BEG \cong \triangle CEG, \triangle CDG \cong \triangle ADG, \triangle AFG \cong \triangle BFG$. Hence, $AG = BG = CG$ is the radius (the *circumradius*) of the circle that passes through A, B, C.

Problem 5.5 Suppose $\triangle ABC$ is a right triangle with $\angle C = 90°$ and $AC = 5, BC = 12$. Let $\overline{AD}$ be an angle bisector. Find the area of $\triangle ABD$.

Answer

$\dfrac{65}{3}$.

Solution

Note using the Pythagorean Theorem, $AB = 13$. If $x = DC$, then $BD = 12 - x$. Using the Angle Bisector Theorem, $\dfrac{13}{5} = \dfrac{AB}{AC} = \dfrac{BD}{DC} = \dfrac{12 - x}{x}$. Solving for x we have $x = \frac{10}{3}$ and hence $12 - x = \frac{26}{3}$. Therefore, the area of $\triangle ABD$ is $\frac{65}{3}$.

Problem 5.6 Viviani's Theorem

(a) Recall a more general version of the problem we did earlier: Let triangle ABC be a equilateral triangle with side length s. Let point P be an arbitrary point in the interior of triangle ABC, D be point on $\overline{BC}$, E be point on $\overline{CA}$, and F be point on $\overline{AB}$, so that $\overline{PD} \| \overline{AB}$, $\overline{PE} \| \overline{BC}$, and $\overline{PF} \| \overline{AC}$. Find the value of $PD + PE + PF$.

Answer

s.

Solution

Note that $PDCE, PFBD, PEAF$ are all isosceles trapezoids. By the isosceles trapezoid problem, $PD + PE = CD$. Further, $PF = DB$, so $PD + PE + PF = CD + DB = CB = s$.

(b) (Viviani's Theorem) Let triangle ABC be an equilateral triangle with side length s and therefore heights/altitudes of length $h = \dfrac{s\sqrt{3}}{2}$. Let P be an arbitrary point in the interior of the triangle. Show that the sum of the distances from P to all three sides is equal to h.

Solution

Start with the drawing from the previous problem. If we extend the lines we divide $\triangle ABC$ into three equilateral triangles and three parallelograms. Let $\overline{PG}, \overline{PH}, \overline{PI}$ be altitudes of these equilateral triangles (with G, H, I on respectively $\overline{BC}, \overline{CA}, \overline{AB}$). If $\overline{BQ}$ is an altitude of $\triangle ABC$, note we have $\triangle BQA \sim \triangle PGD \sim \triangle PHE \sim PIF$. Hence,

$$\frac{PD}{PG} = \frac{PE}{PH} = \frac{PF}{PI} = \frac{BA}{BQ} = \frac{s}{h}$$

and using the previous result we have $PD + PE + PF = BA = s$ so $PG + PH + PI = BQ = h$.

Problem 5.7 Prove that the coordinates of the centroid of $\triangle ABC$ are given by

$$\left(\frac{x_A + x_B + x_C}{3}, \frac{y_A + y_B + y_C}{3} \right).$$

Further prove that the centroid divides the medians into ratios of $2 : 1$.

Solution

We prove the result for the median from A (as the others are identical). The median from A starts at A and goes to the midpoint of $\overline{BC}$ which is $\left(\frac{x_B + x_C}{2}, \frac{y_B + y_C}{2} \right)$. Using the generalized midpoints formula, the point $\frac{2}{3}$ of the way from A to this midpoint is

$$\left(\left(1 - \frac{2}{3}\right) x_A + \frac{2}{3} \cdot \frac{x_B + x_C}{2}, \left(1 - \frac{2}{3}\right) y_A + \frac{2}{3} \cdot \frac{y_B + y_C}{2} \right)$$
$$= \left(\frac{x_A + x_B + x_C}{3}, \frac{y_A + y_B + y_C}{3} \right)$$

which is the centroid as needed.

Problem 5.8 Let quadrilateral $ABCD$ be in the diagram below, AE and AF are angle bisectors.

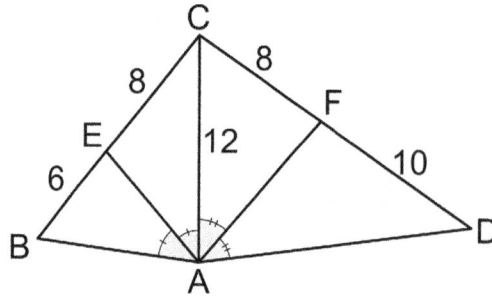

Find the perimeter of *ABCD*.

Answer

56.

Solution

Using the angle bisector theorem we have $AB : AC = BE : EC = 3 : 4$ so $AB = 9$.
Similarly we find $AD = 15$. Hence the perimeter is $9 + 14 + 18 + 15 = 56$.

Problem 5.9 Suppose three congruent circles are all tangent inside a larger circle.

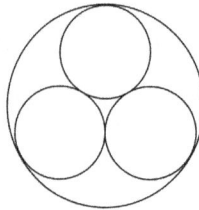

If the bottom two circles have centers $(2,0)$ and $(4,0)$ find the equation of the large circle.

Answer

$$(x-3)^2 + \left(y - \frac{\sqrt{3}}{3}\right)^2 = \left(1 + \frac{2\sqrt{3}}{3}\right)^2.$$

Solution

Use the following diagram to help:

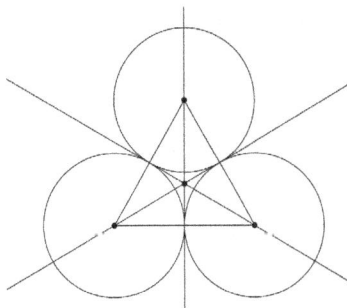

Since the bottom two circles have centers $A = (2,0), B = (4,0)$ their tangent point must be $(3,0)$ and hence both circles have a radius of 1. Hence, the triangle formed by the centers of the three small congruent triangles must be an equilateral triangle with side length 2. From this we can calculate the center of the third smaller triangle to be $C = (3, \sqrt{3})$. As all the centers (orthocenter, centroid, incenter, circumcenter) of an equilateral triangle are the same, we can use our centroid formula from earlier to calculate the point $D = (3, \sqrt{3}/3)$ to be the orthocenter of $\triangle ABC$. The length of $\overline{AD}$ is $2\sqrt{3}/3$, and hence the radius of the large circle is $2\sqrt{3}/3 + 1$.

Problem 5.10 (Euler Line) Let ABC be a triangle with centroid G and circumcenter O.

(a) Let H be on ray $\overrightarrow{OG}$ so that $OG : GH = 1 : 2$. Prove that $\overline{AH} \perp \overline{BC}$.

Solution

Let D be the midpoint of $\overline{BC}$. By contruction of H, we have that $\triangle GDO \sim GAH$ by SAS. Hence $\angle GDO = \angle GAH$ and hence $\overline{DO} \parallel \overline{AH}$ and hence $AH \perp \overline{BC}$ as needed.

(b) Prove that the centroid, circumcenter, and orthocenter of $\triangle ABC$ are collinear. If this line is unique it is called the Euler line.

Solution

Note by part (a) H is on the altitude from A. Since the centroid and circumcenter are the intersections of the medians and perpendicular bisectors respectively, repeating part (a) shows that H is on the altitudes from B and C. Hence H must be the orthocenter.

6   Solutions to Chapter 6 Examples

Problem 6.1 Suppose you have a trapezoid $ABCD$ with $\overline{AB}$ parallel to $\overline{CD}$. Let E be the intersection of the diagonals. Suppose $AB = 10, CD = 15$ and $\triangle ADE$ has area 24. Find the area of $ABCD$.

Answer

100.

Solution

Since $\overline{AB}\|\overline{CD}$, $\triangle ABE \sim \triangle CDE$, with ratio of corresponding sides $10 : 15 = 2 : 3$. Hence $DE : EB = 3 : 2$ and since $\triangle AED, \triangle AEB$ share the same height from A, $[AED] : [AEB] = 3 : 2$ so $[AEB] = 24 \cdot \frac{2}{3} = 16$. We can use a similar argument to get $[DEC] = 36$ and $[BEC] = 24$. Hence the total area is 100.

Problem 6.2 Let $\overline{AM}$ be a median of $\triangle ABC$, and D be a point on $\overline{MC}$, and E be a point on $\overline{AB}$, such that $\overline{ME} \parallel \overline{AD}$. Show that $[BDE] = [AEDC]$.

Solution

Let O be the intersection of $\overline{AM}$ and $\overline{DE}$. Then $[AOE] = [MOD]$ because $\overline{ME} \parallel \overline{AD}$. Also $[ABM] = [ACM]$ because M is the midpoint of $\overline{BC}$, thus $[BMOE] = [AODC]$. Finally, $[BDE] = [AEDC]$.

Problem 6.3 Suppose the altitudes of a triangle are in ratio $2 : 2 : 3$ and the triangle has a perimeter of 24. Find the area of the triangle.

Answer

$18\sqrt{2}$.

Solution

As in an earlier problem, we use the fact that we can calculate the area using all of the altitudes. Suppose the altitudes (from A, B, C respectively) are h_A, h_B, h_C with $h_A : h_B : h_C = 2 : 2 : 3$. If a, b, c are the opposite sides, we then have (after cancelling $1/2$): $a \cdot h_A = b \cdot h_B = c \cdot h_C$. We therefore have that $b : a = h_A : h_B = 2 : 2$ and $c : b = h_B : h_C = 3 : 2$. Hence, the sides are in ratio $c : b : a = 2 : 3 : 3$. If the perimeter of the

---

triangle is 24, this means the sides of the triangles are $6, 9, 9$. Using Heron's formula, the area is thus, $18\sqrt{2}$.

Problem 6.4 Let G be the centroid of $\triangle ABC$, and $AG = 3, BG = 4, CG = 5$, find $[ABC]$.

Answer

18.

Solution

Say the medians are $\overline{AD}, \overline{BE}, \overline{CF}$, which divide $\triangle ABC$ into six equal area triangles. Extend $\overline{BE}$ 2 units further, giving a point H. Note $AGCH$ is a parallelogram, so $CH = 3$. Hence, $\triangle GCH$ is a right triangle with area 6. Note that $[GCH] = [ACG] = [ABC]/3$ and thus $[ABC] = 18$.

Problem 6.5 Let M be the intersection of line segments $\overline{AB}$ and $\overline{PQ}$. Show that $\dfrac{[PAB]}{[QAB]} = \dfrac{PM}{QM}$.

Solution

Since $\dfrac{[APM]}{[AQM]} = \dfrac{PM}{QM}$ (the two triangles have the same height), and $\dfrac{[BPM]}{[BQM]} = \dfrac{PM}{QM}$ (similar reason), the result is obtained (remember if $a : b = c : d$ then $a + c : b + d = a : b$).

Problem 6.6 (2002 AMC 12A #22) Triangle ABC is a right triangle with $\angle ACB$ as its right angle, $m\angle ABC = 60°$, and $AB = 10$. Let P be randomly chosen inside $\triangle ABC$, and extend $\overline{BP}$ to meet $\overline{AC}$ at D. What is the probability that $BD > 5\sqrt{2}$?

Answer

$\dfrac{3 - \sqrt{3}}{3}$.

Solution

Triangle ABC is a 30-60-90 triangle, and $BC = 5$, $AC = 5\sqrt{3}$. Let E be the point on $\overline{AC}$ such that $CE = 5$. Then $BE = 5\sqrt{2}$. The desired probability is for P to fall in the triangle ABE, and equals the ratio $\dfrac{ABE}{ABC} = \dfrac{AE}{AC} = \dfrac{5\sqrt{3} - 5}{5\sqrt{3}} = \dfrac{3 - \sqrt{3}}{3}$.

Problem 6.7 Suppose a dodecagon is in a square as in the diagram below (the vertices of the dodecagon on the square are the midpoints of the sides):

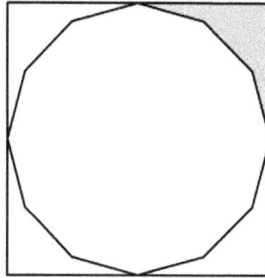

If the area of the square is 4, what is the area of the shaded region?

Answer

$\dfrac{1}{4}$.

Solution

Consider the following diagram

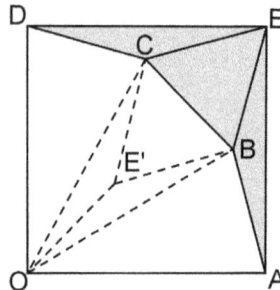

where $OABC$ is a quarter of the dodecagon and E' is E reflected across $\overline{BC}$. $\triangle OAB, \triangle BOC, \triangle COD$ are all congruent isosceles triangles with angles $30°, 75°, 75°$. Clearly $\triangle BCE \cong \triangle BCE'$ and calculating angles we see $\triangle ABE, \triangle ECD, \triangle OE'B, \triangle CE'O$ are all congruent isosceles triangles with angles $15°, 15°, 150°$. Hence the shaded region has area $[OAED]/4$ which is $1/16$th of the entire square. Hence the shaded region has area $1/4$.

Problem 6.8 Let $ABCDE$ be a convex pentagon. Suppose further that the triangle cut off by each diagonal has area 1. What is the area of the full pentagon $ABCDE$?

Answer

$$\frac{5+\sqrt{5}}{2}.$$

Solution

We are given that $[ABC] = [BCD] = [CDE] = [DEA] = [EAB] = 1$. Note $\triangle EAB$ and $\triangle ABC$ share the base AB so they have the same heights. This implies that $\overline{AB}$ is parallel to $\overline{CE}$. A similar argument gives $\overline{BC} \parallel \overline{AD}$, $\overline{CD} \parallel \overline{BE}$, etc. Thus, we get the following type of diagram.

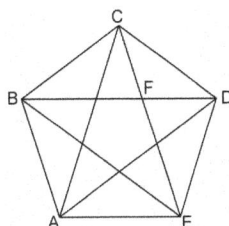

Set $[BCF] = x$. Then $[CFD] = 1 - x$. Using these triangles we have

$$\frac{BF}{FD} = \frac{[BCF]}{[CFD]} = \frac{x}{1-x}.$$

Also, note $[BFE] = [ABE] = 1$ as $ABFE$ is a parallelogram. Hence using triangles $\triangle BFE$ and $\triangle DFE$,

$$\frac{BF}{FD} = \frac{[BFE]}{[DFE]} = \frac{1}{x}.$$

Hence $\dfrac{x}{1-x} = \dfrac{1}{x}$ so we can solve for $x = (\sqrt{5} - 1)/2$ (ignoring the negative root). Lastly,

$$[ABCDE] = [ABE] + [BFE] + [BCF] + [CDE] = 1 + 1 + x + 1 = 3 + x = \frac{5+\sqrt{5}}{2}.$$

Problem 6.9 Start with trapezoid $ABCD$. Extend $\overline{AD}, \overline{BC}$ to meet at O as in the diagram below.

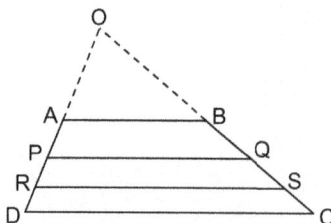

Suppose $AD = 1, OA = 1$. Construct P, Q, R, S such that $\overline{AB} \parallel \overline{PQ} \parallel \overline{RS}$ and $[ABQP] = [QSRP] = [SCDR]$. What are AP, PR, RD?

Answer

$AP = \sqrt{2} - 1, PR = \sqrt{3} - \sqrt{2}, RD = 2 - \sqrt{3}$.

Solution

Note that $\triangle OAB \sim \triangle OPQ \sim \triangle ORS \sim \triangle ODC$. Since $OA = AD$ we have $[OAB] = [ABCD]/4$. Hence we want $[OPQ] = [ABCD]/2 = 2[OAB], [ORS] = 3[ABCD]/4 = 3[OAB]$. Thus, $OP = \sqrt{2}, OR = \sqrt{3}$ and hence $AR = \sqrt{2} - 1$, $PR = \sqrt{3} - \sqrt{2}$, $RD = 2 - \sqrt{3}$ as needed.

Problem 6.10 (Kurrah's Theorem) Let $\triangle ABC$ be given. Construct D, E as in the diagram below with $\angle ABC = \angle ADB = \angle BEC$.

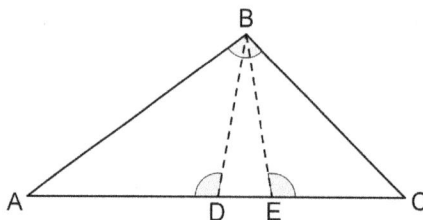

Prove that

$$AB^2 + BC^2 = AC(AD + CE).$$

Solution

Note we have $\triangle ABC \sim \triangle ADB \sim \triangle BEC$. Thus

$$\frac{AB}{AC} = \frac{AD}{AB} \Rightarrow AB^2 = AC \cdot AD$$

and

$$\frac{BC}{AC} = \frac{CE}{BC} \Rightarrow BC^2 = AC \cdot CE.$$

Adding we get $AB^2 + BC^2 = AC(AD + CE)$ as needed.

7 Solutions to Chapter 7 Examples

Problem 7.1 Storing Tires

(a) Suppose two tires, each with radius 1ft rest upright on the ground and touching each other, as pictured below:

How much space is needed horizontally to store the tires?

Answer

4.

Solution

Note the distance between the center of the circles is parallel to the ground, so the total distance needed is the two diameters, which sum to 4ft.

(b) Repeat part (a) with two tires of radius $1, 2$ feet respectively.

Answer

$3 + 2\sqrt{2}$.

Solution

To store the tires using the least amount of horizontal space, we again examine the line $\overline{AB}$ connecting the two centers, as in the diagram below:

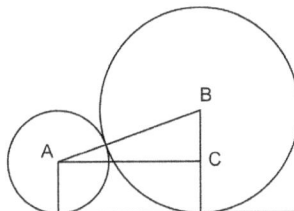

From here we see the horizontal distance in between the centers of the tires is AC, where $\triangle ABC$ is a right triangle with $AB = 1 + 2 = 3, BC = 2 - 1 = 1$. Hence $AC = 2\sqrt{2}$ and the total distance needed is $1 + 2\sqrt{2} + 2$.

Problem 7.2 Suppose you start with a circle of radius 1. Draw another circle of radius 1 with center an arbitrary point on the first circle. Let R denote the region consisting of all points that are inside both circles.

(a) Find the perimeter of R.

Answer

$4\pi/3$.

Solution

Let A, B be the centers of the circle, and C, D be the intersection points of the two circles. Hence the perimeter of R consists of the arc from C to D (containing A) plus the arc from D to C (containing B). Further, as both $\triangle ABC, \triangle ABD$ are equilateral triangles (SSS), each are is seen to be $120°$, or $1/3$ of a circle. Hence the perimeter of R is $2 \cdot \dfrac{1}{3} \cdot 2\pi = \dfrac{4\pi}{3}$.

(b) Find the area of R.

Answer

$2\pi/3 - \sqrt{3}/2$.

Solution

Consider the labeling from above. We then have that the area of R is the sum of the areas of the two sectors minus the two equilateral triangles: $2 \cdot \dfrac{1}{3} \cdot \pi - 2 \cdot \dfrac{\sqrt{3}}{4}$.

Problem 7.3 Inscribing Circles in Sectors

(a) What is the radius of the largest circle that can fit in a quarter circle of radius 1?

Answer

$\sqrt{2} - 1$.

Solution

Consider the following general diagram:

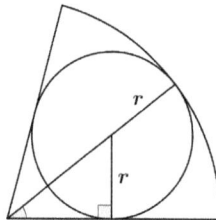

where if the sector is $\theta°$, the marked angle in the right triangle is $\theta/2$.
In this case, $\theta = 90°$, so the triangle is a 45-45-90 triangle with hypotenuse $r\sqrt{2}$. Hence,
$1 = r + r\sqrt{2}$ so solving for r gives $r = \dfrac{1}{1+\sqrt{2}} = \sqrt{2} - 1$.

(b) What if instead you are fitting it into a 60°-sector?

Answer

$1/3$.

Solution

A similar method to part (a) gives the equation $1 = r + 2r$ so $r = 1/3$.

Problem 7.4 Let $\angle APB$ be an inscribed angle on a circle with center O. Prove that $\angle APB$ is half the angular size of arc $\overset{\frown}{AB}$ if:

(a) O lies on $\angle APB$.

Solution

Assume O lies on $\overline{AP}$. Then $\triangle OPB$ is isosceles, so $2\angle BPO = 2\angle BPA = \angle BOA$ as needed.

(b) O lies inside $\angle APB$.

Solution

Let Q be such that $\overline{PQ}$ is a diameter. Note $\triangle BOP$ is isosceles, so $\angle BOQ = 2\angle BPO$. Similarly, $\angle AOQ = 2\angle APO$. Hence, $\angle BPA = \angle BPO + \angle APO = (\angle BOQ + \angle AOQ)/2 = \angle BOA/2$ as needed.

Problem 7.5 Prove that if two chords AC, BD intersect inside a circle at point P then the measure of $\angle APB$ is half the sum of the angular sizes of $\overparen{AB}, \overparen{CD}$.

Solution

Draw $\overline{BC}$. We have $\angle APB = \angle DBC + \angle ACB$ equals have the sum of the angular sizes of $\overparen{CD}, \overparen{AB}$ (as $\angle DBC, \angle ACB$ are inscribed angles).

Problem 7.6 Suppose $\overparen{AB}$ is an arc with angular size $60°$ and CD is a diameter such that if rays $\overrightarrow{BA}, \overrightarrow{DC}$ are extended to intersect at a point E, $\angle AEC = 30$. Find the angular size of arc $\overparen{BD}$.

Answer

$90°$.

Solution

Let the angular measure of $\overparen{BD} = x$. Then the size of $\overparen{AC} = 180 - 60 - x = 120 - x$. Then $\angle AEC = 30 = \dfrac{x - (120 - x)}{2}$ so solving for x gives $x = 90°$.

Problem 7.7 Suppose two perpendicular chords intersect and divide each other in a ratio of $1 : 2$. Find the radius of the circle if each chord is 12in long.

Answer

$2\sqrt{10}$.

Solution

The two chords divide each other in segments of 4 and 8 inches. The perpendicular bisector of each chord goes through the center of the circle, so we can form a right triangle with sides 2 and 6 with hypotenuse r, the radius of the circle. Hence, $r = \sqrt{2^2 + 6^2} = 2\sqrt{10}$.

Problem 7.8 Suppose ω is a circle with radius 6 and center O. Let $\overparen{AB} = 135°$. Let C be on ω such that $\overline{OA} \parallel \overline{BC}$. Find $[OACB]$.

Answer

$18 + 9\sqrt{2}$.

Solution

Let E on $\overline{BC}$ be such that $\overline{OE}$ is a height. As $\overline{OE} \perp \overline{BC}$ and O is the center of the circle, E is the midpoint of $\overline{BC}$. Further, $\triangle OEB$ is an isosceles right triangle. Hence, the trapezoid has bases $6, 6\sqrt{2}$ and height $3\sqrt{2}$, so it has area $\frac{1}{2} \cdot (6 + 6\sqrt{2}) \cdot 3\sqrt{2} = 18 + 9\sqrt{2}$.

Problem 7.9 (AMC 12A 2007 #10) A triangle with side length in the ratio $3 : 4 : 5$ is inscribed in a circle of radius 3. What is the area of the triangle?

Answer

$\dfrac{216}{25}$.

Solution

Let the sides of the triangle have lengths $3x$, $4x$, and $5x$. The triangle is a right triangle, so its hypotenuse is a diameter of the circle. Thus $5x = 2 \cdot 3 = 6$, so $x = 6/5$. The area of the triangle is $\frac{1}{2} \cdot 3x \cdot 4x = \dfrac{216}{25}$.

Problem 7.10 Let $\mathscr{C}_1$ and $\mathscr{C}_2$ be circles defined by

$$(x - 10)^2 + y^2 = 36$$

and

$$(x + 15)^2 + y^2 = 81,$$

respectively. What is the length of the shortest line segment $\overline{PQ}$ that is tangent to $\mathscr{C}_1$ at P and to $\mathscr{C}_2$ at Q?

Answer

20

Solution

The circle $\mathscr{C}_1$'s center is $(10,0)$ and radius 6, and $\mathscr{C}_2$ center $(-15,0)$ and radius 9. Calculate the length of the internal tangent line. This line passes through the origin (why?). There are two triangles of the 3-4-5 side-ratios with actual side lengths $(6,8,10)$ and $(9,12,15)$, and the length of the tangent line is $8+12=20$. The external tangent line is longer: $\sqrt{25^2-(9-6)^2}=\sqrt{616}$.

8   Solutions to Chapter 8 Examples

Problem 8.1 Prove that if $\angle A + \angle C = \angle B + \angle D = 180°$ in quadrilateral $ABCD$, then $ABCD$ is a cyclic quadrilateral .

Solution

Start by circumscribing a circle about $\triangle ABC$. If D is not on the circle, then $\angle B + \angle D$ is either greater or less than $180°$ (depending on whether D is inside or outside the circle.

Problem 8.2 Suppose $ABCD$ is a quadrilateral.

(a) Prove that if $ABCD$ is cyclic, then $\angle ABD = \angle ACD$ (and similarly $\angle BAC = \angle BDC$, etc.).

Solution

$\angle ABD, \angle ACD$ both share the arc $\overset{\frown}{AD}$ (not containing B, C), so they are the same angle.

(b) Prove that if $\angle ABD = \angle ACD$ (or $\angle BAC = \angle BDC$, etc.), then $ABCD$ is cyclic.

Solution

Let ω be the circumcircle of $\triangle ABD$. By assumption $\angle ACD = \angle ABD$ which is half the angular measure of $\overset{\frown}{AD}$. If C is were not on ω, $\angle ABD$ would either be greater or smaller than half the angular measure of $\overset{\frown}{AD}$ (respectively if D is inside or outside ω), a contradiction.

Problem 8.3 Prove the Power of a Point formula for $\overline{AB}$ and $\overline{CD}$ intersecting inside the circle.

Solution

If $\overline{AB}$ and $\overline{CD}$ intersect at E, we want to show that $AE \cdot BE = CE \cdot DE$.

Since they share arcs, we have that $\angle ADC = \angle ABC$ and $\angle BAD = \angle BCD$. Therefore $\triangle ADE \sim \triangle CBE$ (using AA). Hence $AE/DE = CE/BE$ or, after cross-multiplying, $AE \cdot BE = CE \cdot DE$ as needed.

Problem 8.4 Prove Ptolemy's theorem, using the following diagram as guidance,

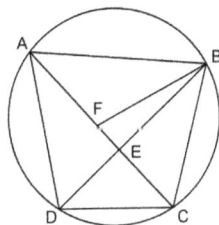

where F is such that $\angle ABF = \angle CBD$.

Solution

We constructed F such that $\triangle ABF \sim \triangle DBC$, and we also have $\triangle ABD \sim FBC$ (make sure you can fill in the details here). Therefore, we have that $AF \cdot BD = AB \cdot CD, CF \cdot BD = BC \cdot DA$. Adding these equalities and simplifying gives $AC \cdot BD = AB \cdot CD + AD \cdot BC$. as needed.

Problem 8.5 Suppose chords $\overline{AB}, \overline{CD}$ intersect at E, such that $AE : EB = 1 : 3$ and $CE : ED = 1 : 12$. Find the ratio of $AB : CD$.

Answer

$8/13$.

Solution

Let $AE = x, CE = y$, so $EB = 3x, ED = 12y$. By Power of a Point, $3x^2 = 12y^2$, so solving for x/y we get $x/y = 2$. Hence, $\dfrac{AB}{CD} = \dfrac{4x}{13y} = \dfrac{4}{13} \cdot 2 = \dfrac{8}{13}$.

Problem 8.6 (AHSME 1999 #21) A circle is circumscribed about a triangle with sides 20, 21, and 29, thus dividing the interior of the circle into four regions. Let A, B, and C denote the areas of the non-triangular regions, with C being the largest. Compute $C - (A + B)$.

Answer

210.

Solution

Since 20, 21, and 29 form a Pythagorean triple, the triangle is a right triangle, and the side with length 29 is the diameter. It is easy to see that the result of $C - (A + B)$ is exactly the area of $\triangle ABC$.

Problem 8.7 Let A, B, C, D be four points, arranged in clockwise order, on circle ω. Segments AC and BD intersect at P. Given that $AB = 3, BP = 4, PA = 5, PC = 6$, find the radius of circle ω.

Answer

$\sqrt{565}/4$.

Solution

We find the diameter. From the given lengths it is easy to see that $\triangle ABP$ is a right triangle, and $\angle ABP = 90°$. So AD is the diameter. By Power of a Point, $AP \cdot PC = BP \cdot PD$, so $PD = 5 \cdot 6/4 = 15/2$. In right triangle ABD, apply the Pythagorean theorem to get $AD = \sqrt{AB^2 + AD^2} = \sqrt{565}/2$. Thus the radius is $\sqrt{565}/4$.

Problem 8.8 Suppose we have a rectangle $ABCD$ with $AB = 8$, $BC = 12$. Inscribe a circle in the rectangle so that it touches sides $\overline{AB}, \overline{BC}, \overline{AD}$. Let M be the midpoint of $\overline{AB}$. Call $E \neq M$ the intersection of $\overline{MD}$ with the circle. Find DE.

Answer

$8\sqrt{10}/5$.

Solution

Let F be such that $\overline{AD}$ is tangent to the circle at F. Since $AB = 8$, the radius of the circle is 4. Hence, $AM = 4$ and $DF = 12 - 4 = 8$. By Power of a Point, $DF^2 = DM \cdot DE$. As $DM = \sqrt{4^2 + 12^2} = 4\sqrt{10}$, $DE = \dfrac{8^2}{4\sqrt{10}} = \dfrac{8\sqrt{10}}{5}$.

Problem 8.9 Given right triangle ABC, construct semicircles on the three sides as shown.

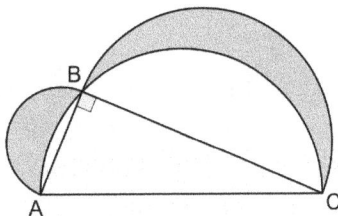

Given that $AB = 5, BC = 12$. Find the sum of the areas of the shaded regions.

Answer

30

Solution

The sum of the areas of the two smaller semicircles is $\frac{1}{2}\left(\frac{\pi}{4}(AB^2 + BC^2)\right) = \frac{\pi}{8}(AB^2 + BC^2) = \frac{\pi}{8}AC^2$, which is exactly the area of the large semicircle. Subtracting the common regions from both sides, the remaining regions are the shaded regions on the side of two smaller semicircles, and the right triangle ABC on the large semicircle. So the sum of the areas of the shaded regions equals the area of the triangle The area $[ABC] = \frac{1}{2} \cdot 5 \cdot 12 = 30$. So the answer is 30.

Problem 8.10 In $\triangle ABC$, $AB = 37, AC = 58$. Use A as center and AB as radius, draw a circle to intersect $\overline{BC}$ at D where D is between B and C.

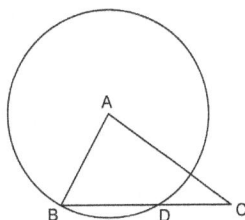

Given that the lengths of $\overline{BD}$ and $\overline{DC}$ are both integers, compute BC.

Answer

57

Solution

Let E be the intersection of $\overline{AC}$ and the circle. Extend $\overline{CA}$ to intersect the circle at F. So $AE = AF = 37$, $CF = 58 + 37 = 95$, and $CE = 58 - 37 = 21$. By Power of a Point, $CB \cdot CD = CF \cdot CE = 95 \times 21 = 1995$. Since the lengths of $\overline{BD}$ and $\overline{DC}$ are integers, so is the length of $\overline{BC}$. Since $\overline{CE}$ goes through A, the center of the circle, $CF > CB$. Since $1995 = 3 \times 5 \times 7 \times 19$, the only way that 1995 is expressed as the product of two factors, and both factors are less than 95, is $1995 = 57 \times 35$. Thus $BC = 57$ and $CD = 35$. Hence 57 is the final answer.

9 Solutions to Chapter 9 Examples

Problem 9.1 Find the volume of the largest sphere that can fit inside a cone of radius 1 and height $\sqrt{3}$.

Answer

$\dfrac{4\sqrt{3}}{27}\pi.$

Solution

Using a 30-60-90 triangle we the radius of the sphere is equal to $1/\sqrt{3}$.

Problem 9.2 Show that the surface area of a regular tetrahedron with side length a is $a^2\sqrt{3}$.

Solution

A regular tetrahedron is made up of 4 equilateral triangles. Each of these has area $a^2\sqrt{3}/4$ so the total surface area is $a^2\sqrt{3}$.

Problem 9.3 Show that the volume of a regular tetrahedron with side length a is $\dfrac{a^3\sqrt{2}}{12}.$

Solution

Note the result follows if we show a regular tetrahedron (which is a triangular pyramid) has height $a\sqrt{2}/\sqrt{3}$. Recall the apex is above the incenter of the base. For an equilateral triangle, the incenter is $a/\sqrt{3}$ from the vertex. This leads to the following diagram:

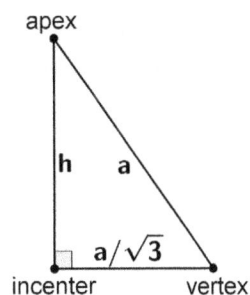

Hence, solving for h, we get $h = a\sqrt{2/3}$ as needed.

Problem 9.4 (2010 AMC 10A #20) Suppose a bored bee lives on a cube with side length 1. For "fun" he decides to visit every vertex of the cube, each exactly once, starting and ending at the same vertex. It will travel from one vertex to another using straight lines (either crawling or flying). Give an example of a path that uses the maximum distance and find this distance.

Answer

$4\sqrt{2} + 4\sqrt{3}$.

Solution

The distance between opposite corners of the cube is $\sqrt{3}$ and the distance opposite corners of a square face is $\sqrt{2}$. Note any path has 8 stops, and it is possible to travel only using the above distances, 4 times each. This is optimal, as there are only 4 pairs of opposite corners (and the bee can only use each once). An example path is given below:

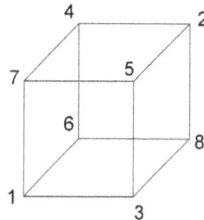

(The bee travels from 1 to 2 to 3 ... to 8 to 1.)

Problem 9.5 Four identical balls (spheres), each of radius 1 in, are glued to the ground so that their centers form the vertices of a square with side length 2 in. Suppose you rest a fifth identical ball on the four balls (so the fifth ball is a sphere externally tangent to the other spheres). How far does this ball rest off the ground?

Answer

$\sqrt{2}$.

Solution

Using symmetry we look at spheres with centers across the diagonal:

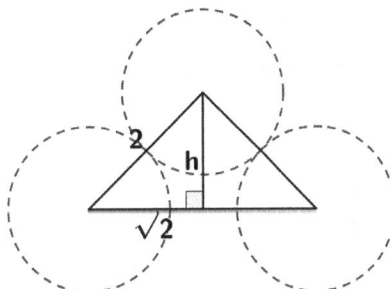

Let $h+1$ denote the height of the center of the fifth sphere off the ground. Setting up right triangles gives us $2^2 = 2 + h^2$, so solving for h gives $\sqrt{2}$.

Problem 9.6 Suppose you pick 4 vertices of a cube to form a tetrahedron.

(a) How many different (non-congruent) tetrahedra are possible? Are any of them regular tetrahedra?

Answer

4

Solution

First assume, 3 of the vertices are contained in the same square face. Label the cube $ABCD - A'B'C'D'$ so that the vertices chosen are A, B, C, and one of A', B', C', D'. Note $ABC - B' \cong ABC - C'$, so this leads to 3 different tetrahedron, none of which are regular.

The only other possibility (up to congruence) is (again using the labelling $ABCD - A'B'C'D'$) $AC - B'D'$. Note this tetrahedron *is* regular.

(b) Find the volumes for each of the possibilities in (a) if the cube has volume 1.

Answer

1/6 or 1/3

Solution

If 3 vertices are contained in the same square face, then we have a tetrahedron (which is a triangular pyramid) with base of area $1/2$ and height of 1. Thus, the volume is $\frac{1}{3} \cdot \frac{1}{2} \cdot 1 = \frac{1}{6}.$

Now suppose we have tetrahedron $AC - B'D'$. Note this tetrahedron divides the rest of the cube into 4 congruent tetrahedra, which we just saw have volume $\frac{1}{6}$. Hence, tetrahedron $AC - B'D'$ has volume $\frac{1}{3}$.

Problem 9.7 Suppose you have a unit cube. Pick two opposite corners. In each corner, form a tetrahedron using the corner and the three adjacent vertices. Remove these two tetrahedra and call the resulting polyhedron $\mathscr{S}$.

(a) How many vertices, edges, and faces does the resulting polyhedron have? Describe the faces.

Answer

6 vertices, 12 edges, and 8 faces.

Solution

Note we are removing 2 vertices, not changing the number of edges, and adding 2 faces. All of the faces are triangles.

(b) Find the volume of $\mathscr{S}$.

Answer

2/3.

Solution

Note each of the tetrahedra removed have base area $1/2$ and height 1, hence has volume $1/6$. Hence $\mathscr{S}$ has volume $1 - 2 \cdot \frac{1}{6} = \frac{2}{3}$.

Problem 9.8 Suppose you have a sphere of radius 1. That is the side length of the largest regular tetrahedron you can fit (inscribe) inside the sphere?

Answer

$\dfrac{2\sqrt{6}}{3}$

Solution

Recall that picking vertices from a cube with side length s we can form a regular tetrahedron with side length $s\sqrt{2}$. Hence we want to know the largest cube we can fit inside a sphere of radius 1. We know that the diagonal of a cube with side length s is $s\sqrt{3}$ which must be the diameter of the sphere. Hence

$$s\sqrt{3} = 2 \Rightarrow s = \frac{2\sqrt{3}}{3}.$$

Finally, the side length of the tetrahedron is

$$\frac{2\sqrt{3}}{3} \cdot \sqrt{2} = \frac{2\sqrt{6}}{3}.$$

Problem 9.9 An obtuse triangle with dimensions 9, 10, and 17 is rotated about the smallest side so that it creates a three-dimensional solid shown below.

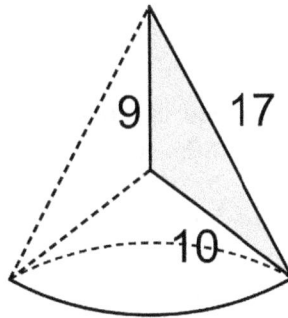

Determine the surface area of the solid. Use $\pi = 3.14$ and round your answer to the nearest tenth if necessary.

Answer

678.2

Solution

Note that in the figure above, there are two cones sharing the same circular base. Let r be the radius of the cones and let h be the height of the smaller cone. Therefore, h

and r satisfy $h^2 + r^2 = 10^2$ and $(h+9)^2 + r^2 = 17^2$. If you recall Pythagorean triples, $h = 6$ and $r = 8$ yields a $6 - 8 - 10$ and $8 - 15 - 17$ Pythagorean triples. Therefore, the surface area of the new figure is

$$\pi(8 \times 17) + \pi(8 \times 10) = 216\pi \approx 678.2.$$

Problem 9.10 Suppose you start with a right cone and cut off the top of the cone with a plane parallel to the base. The resulting solid, called a *frustrum* has two circular "bases", say with radii R and r (with $R > r$), and height h. (Hence from the side the frustrum looks like a trapezoid with bases R, r and height H.)

(a) Show that the volume of a frustrum is $\dfrac{\pi H}{3}(R^2 + Rr + r^2)$.

Solution

Let $H + h$ denote the height of the original cone. Clearly the volume of the frustrum is $\pi R^2(H+h)/3 - \pi r^2 h/3$. Using similar triangles we have $\dfrac{h}{r} = \dfrac{h+H}{R}$ so $h = \dfrac{rH}{R-r}$. Simplifying gives the above result.

(b) Suppose a sphere can be inscribed in a frustrum with base radii r, R such that the sphere is tangent to the two bases and the side. Find the radius of such a sphere in terms of r, R.

Answer

$\sqrt{rR}$.

Solution

Let s denote the radius of the sphere. We have the following diagram:

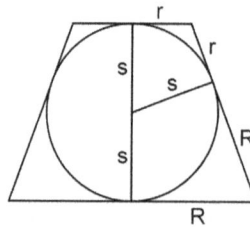

Therefore, $H = 2s$. Further, dropping a height from an endpoint of the top base yields a right triangle, so $(R+r)^2 = (2s)^2 + (R-r)^2$. Solving for s gives $s = \sqrt{rR}$.